D1187297

GUN BARONS

ALSO BY JOHN BAINBRIDGE, JR.

American Gunfight: The Plot to Kill Harry Truman and the Shoot-out That Stopped It (with Stephen Hunter)

This book is dedicated to Deborah and Howard Chasanow, who by their examples showed me that high intellect, hard work, honor, and kindness can reside together comfortably in the same person.

CONTENTS

GUN BARONS

The Weapons That Transformed America and the Men Who Invented Them

• • • • • • • • • • • • •

JOHN BAINBRIDGE, JR.

ST. MARTIN'S PRESS
NEW YORK

First published in the United States by St. Martin's Press, an imprint of St. Martin's Publishing Group

www.stmartins.com

Designed by Omar Chapa

The Library of Congress Cataloging-in-Publication Data is available upon request.

ISBN 978-1-250-26686-6 (hardcover)
ISBN 978-1-250-26687-3 (ebook)

First Edition: 2022

10 9 8 7 6 5 4 3 2 1

CAST OF CHARACTERS

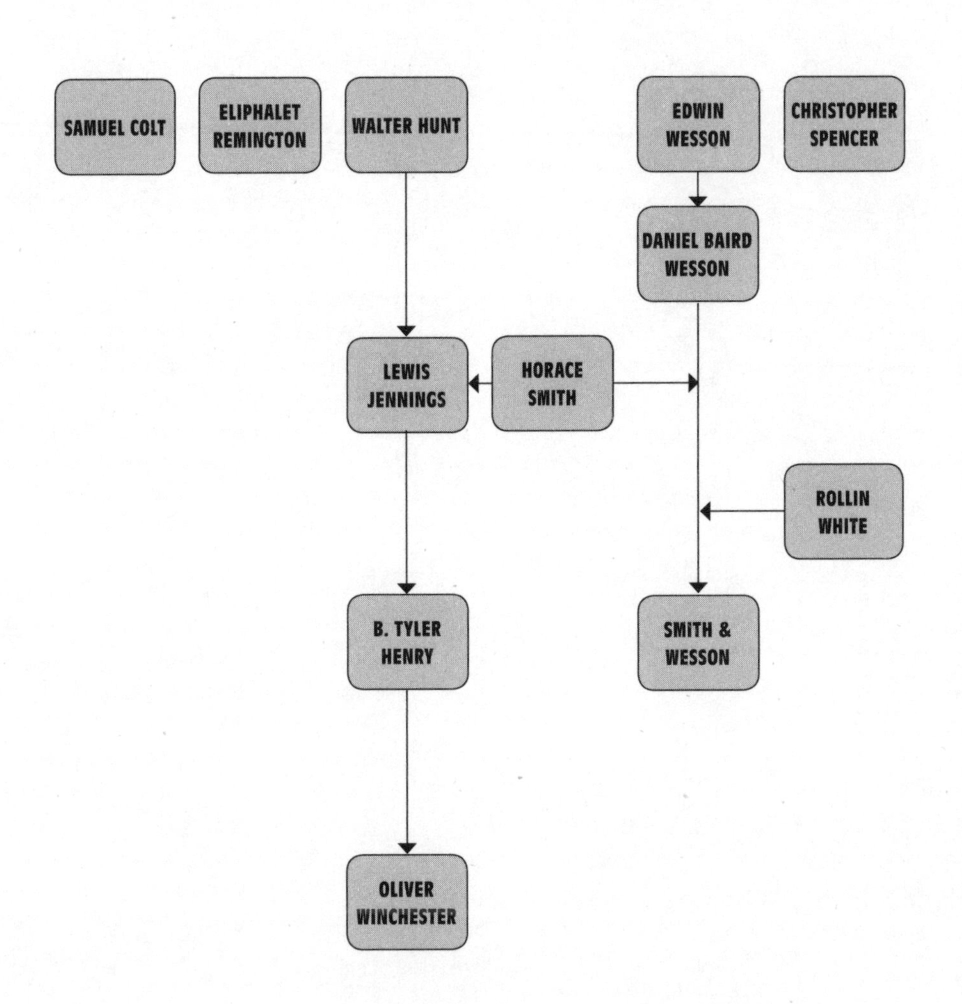

GUN
BARONS

INTRODUCTION

Americans love their guns. Hate them, too.

They are on our hips, in our bedside drawers, welded to our psyche and our language. Our conversations are peppered with gun metaphors: We set our sights, take a long shot, look for a silver bullet, are straight shooters, shoot from the hip, and go off half-cocked. We ride shotgun, sweat bullets, and keep our powder dry. We stick to our guns.

They dwell in our soul. Before its founding the United States was a land where individuals owned and used guns, which became badges of civic responsibility and manhood, summoned in defense of self and community. "Gun owning was so common in colonial America," two legal researchers wrote, "that any claim that eighteenth-century America did not have a 'gun culture' is implausible, just as one could not plausibly claim that early Americans did not have a culture of reading or wearing clothes." In the arms-bearing society of young New England, only the truly poor could not acquire weaponry. And the right to hunt belonged to all, not just to the landed aristocracy; this was not so in old Europe.

The American Revolution fused fiery notions of liberty with ownership of guns. Enlistment posters designed to entice men into becoming soldiers willing to challenge the world's greatest empire fed a growing spirit of independence. In the decades after the United States was born, a mix of national pride, historic reality, and abundant myth cemented gun possession as part of the American character that remained virtually unchallenged through the years following World War II, when entertainment media celebrated the glamorous, well-armed cowboy and stirred millions of children, mostly boys, to carry toy six-guns in hip-hugging holsters.

As long as human beings have used tools, weapons have been those of foremost importance. They have provided food and protection since the formation of the earliest social units. For centuries firearms have been the most effective weapons individuals can wield. Guns have implemented both the highest and the basest goals of humanity—to enforce or defy the law, to defend or acquire territory and treasure, and to liberate or enslave.

Today notions of gun ownership have changed. The frontier is no longer. An armed citizenry is seen by many as not only unnecessary to the nation's survival but a threat to domestic peace. Now guns are at the root of an increasingly fierce debate over who we are as a country and what we believe. They galvanize and polarize us. While they remain touchstones of freedom to some, to others they are objects of loathing. Some museum curators refused to add modern firearms to their displays. "[T]hey have bad karma," explained Dorothy Globus, curator at the Museum of Arts and Design in New York. Arthur Drexler, head of the Museum of Modern Art's Architecture and Design Department, was more explicit. Shortly before his retirement in 1986, he wrote: "Deadly weapons are among the most fascinating and well-designed artifacts of our time, but their beauty can be cherished only by those for whom aesthetic pleasure is divorced from the value of life—a mode of perception the arts are not meant to encourage."

Love them or hate them, guns are out there in the United States, perhaps four hundred million of them in private hands. Many bear the names Colt, Winchester, Remington, and Smith & Wesson, which have become synonymous with American guns. But who were the men behind these names? They were more than just keen inventors and wily businessmen. They were among the founding fathers of American industry. They were visionaries, inspired by the pioneering spirit of their young nation, who ushered in an era of rapid-fire weaponry deadlier than the world had ever seen. By creating what some in the twenty-first century might call the "assault weapons" of their day, they also helped reunite a country divided while contributing to the bloodshed that reunification required. And they nurtured in the general public the seed of gun devotion—some would say obsession—that would divide the country again more than a century later. They drew on a unique brand of American individualism—part fiction, part reality—that promoted their wares to both soldier and Everyman while forming the foundation of modern gun culture. In the process these larger-than-life individuals who peddled their names as well as their products furthered a legacy of citizens venerating personal weaponry on the altar of their being, with liberty forged into the metal of the guns they possess.

Unlike their predecessors—gun craftsmen who toiled in small shops for comparatively meager rewards—these men became industrial titans. They were the first Americans whose vast fortunes came from making firearms. Even the innovators whose creations furthered the evolution of firepower—such as John Hall, creator of a breech-loading flintlock rifle, and Simeon North, a pistol maker who designed one of the first milling machines and contributed to the development of machine-made interchangeable parts—are all but forgotten except by firearms historians. Not so with Colt, Winchester, Smith, Wesson, and Remington. Those names resound today, infused with the romance, mystique, and legend of the American gun.

These gun barons did not rise to prominence alone. Surrounding them were people who abetted their ambitions, financed their ideas, set the stage for their successes, celebrated their inventions, or challenged them for supremacy. Yet many names in this supporting cast have drifted into obscurity. Few today know who Rollin White was, though an idea he patented allowed Smith & Wesson to become dominant gunmakers. Walter Hunt's concept for a repeating gun was the ancestor of the Winchester lever-action rifle, though he could never put his into production. And while irrepressible Yankee inventor Christopher Miner Spencer's repeating rifle successfully challenged Winchester's in the marketplace and contributed to Union victory in the Civil War, the company bearing his name disappeared during peacetime, its name relegated to the past. Without Texas Ranger Samuel H. Walker's help in improving a revolver enough so that the US government bought it for combat service in the Mexican-American War, Samuel Colt might not have risen from the failure he had earned.

The founding fathers of American gun empires lived at the right time. Free of tyranny and drunk on possibility, the United States barreled into the mid-nineteenth century with a sense of conquest and creativity. On individual farms, in urban alleys, and on fresh trails leading westward, bold streaks of innovation ran through the entire young nation. It was an age of inventors, of tinkerers, of risk-takers. Thanks to Cyrus McCormick's reaper, agriculture became more efficient and more profitable. In the South, Eli Whitney's cotton gin enhanced plantation wealth and fed the demand for slaves. Samuel F. B. Morse's telegraph made possible instant communication over great distance, a precious asset for a country poised to sprawl over a continent. The Erie Canal, by connecting the Great Lakes to the Hudson River, pulled the frontier closer to the Eastern seaboard. And in New England, Charles Goodyear, a hardware merchant who had spent time in jail for debts, created a process to keep rubber from melting in hot weather; his "vulcanized" product would be found in waterproof clothing, shoes, balls, life jackets, and eventually automobile tires, though he would die penniless. Canals and improved roadways and eventually railroads smoothed travel throughout the Northeast, juicing commerce and widening the marketplace. Agriculture remained the country's economic and moral backbone, yet more and more people abandoned family farms for jobs in cities and factory towns. Their lives were similar to their rapidly evolving nation: youthful, bold, ambitious, full of hubris, on the move. "Can-do" was a cultural trait. American inventiveness surfaced everywhere.

The American Revolution gave us freedom—most of us, at least. The Industrial Revolution, in the following century, gave us power. Gunmakers, part of that second revolution, changed how Americans fought. Sometimes gunmakers cooperated with each other, but often they competed and occasionally battled in court. War favored them all—including war in Europe. A bewildering assortment of objects, including guns, swamped the US Patent Office in applications for innovative contrivances of all kinds—some marketable, many not, but each submitted in the hope of creating valuable technology while enriching its creator. New ideas were the lifeblood of a young republic enthralled with a rising sense of what could be. One patent in particular would help further this revolution, in which American genius imbued a handheld marvel of engineering with fearsome, history-making power.

Walter Hunt was an intellectually restless man who embodied the quests and contradictions of his time. Born on a valley farm in the foothills of the Adirondack Mountains twenty years after the country declared its independence, he came of age in this era of fervor for invention and moved to New York City at the age of thirty. Hunt stood out in any crowd. He was six foot three and powerfully built, his country-boy face a ruddy hue. What also stood out was his unquenchable thirst for inventing. His vigorous imagination conceived of such diverse devices as a fountain pen, an inkstand, a nail maker, a street sweeper, an iceboat, a flax spinner, a shirt collar, a lamp, a streetcar bell, and a new way of attaching heels to boots. Hunt also invented the first workable cross-stitch sewing machine, which he chose not to patent, supposedly for fear that it might put seamstresses out of work, a forbearance he would later regret. He even concocted an elixir marketed as "Hunt's Restorative Cordial" to relieve pain, sustain "vital energy," overcome sleeplessness, calm nerves, cure "Physical Prostration from any cause whatever," and relieve "Bowel Complaints in their worst forms." He was among the most prolific patent-seekers of the time.

Hunt had one problem: a habit of selling his patents to support his wife and four children, leaving the real profits to those who acquired the rights to his more useful inventions. One day, perhaps in agitation over a debt he owed a draftsman for patent drawings, Hunt took to fiddling idly with a piece of resilient wire. After coiling it in the middle, he squeezed the two ends together and let them spring out again. Then he fashioned a small clasp at one end to hold the other end, so that the wire would remain bent. Finally, he placed a metal shroud over the ends to keep their points from poking out and injuring people. This simplest of devices, completed in just three hours, led to US Patent No. 6,281. Hunt called his invention a "dress-pin," ideal for fastening clothing. We know it today as the safety pin. Hunt paid his obligation to the draftsman—there would be more debts to come—and eventually sold the revolutionary pin patent for $400, the equivalent of more than $14,000 today.

The safety pin is one of Walter Hunt's legacies, his clearest contribution to the simple technologies that keep modern life together. But he had an equally profound (though less apparent) effect when he channeled his imagination toward firearms. Guns were raw technology, tools of power and independence that inventors spent the nineteenth century altering and perfecting, increasing their accuracy, efficiency, and killing power. Traditionally a gunman had one shot. In 1847 Hunt conceived of a weapon capable of firing many times without reloading, a recurring goal of gun designers for centuries. If it worked right, he stood to reap great wealth, or at least enough money to take care of his family. War with Mexico was under way, and armies always needed the best guns available; the more a soldier could shoot, the more effective he would be in combat.

The genius of Hunt's safety pin was its simplicity. Creating a weapons system capable of shooting multiple times in quick succession, however, was a far greater challenge and an unlikely pursuit for a man from a family of peace-loving Quakers. Standard guns at the time required a shooter to pour gunpowder down the barrel, follow that with a lead ball, and then place a priming cap on a metal nipple near the other end of the barrel for each shot. When struck by a hammer, the primer would send burning fulminate into the powder, which would ignite and then propel the ball out the muzzle. By eliminating these steps, Hunt created a self-contained "Rocket Ball." The gun he invented to shoot this ammunition—Hunt dubbed it the "Volition Repeater"—is an odd-looking weapon. Two ringed levers beneath the long, skinny action allowed the would-be shooter to seat Rocket Balls one at a time into the chamber from a tube under the barrel. A shooter could theoretically repeat this process until all dozen cartridges in the tube are discharged, as long as its cluster of parts moved in concert as intended.

With this would-be firearm, Walter Hunt had made the nation's future but not his own. The Volition Repeater never worked quite right. His model was beset by kinks. Its mechanism was complex, its parts delicate.

Hunt didn't have the money to finance further development, so he followed his old habit. He sold the patent rights to the repeating rifle, leaving others to transform his inspired design into highly marketable arms. Among those who benefited from Walter Hunt's genius were Oliver Winchester, Horace Smith, and Daniel Baird Wesson. None of these gun barons possessed the broadly inventive mind of Walter Hunt, yet all would eclipse Hunt while taking advantage of his pioneering work in weaponry. Their main quest was the same as Hunt's: mass-produced, handheld weapons that an individual could keep shooting without stopping to reload, guns that justified the name "repeaters." It was not a new objective, but its fruition in metal and wood had eluded the most dedicated machinists.

In achieving their goal, the gun barons had a major advantage. Mass production had already begun in the United States, and the country's biggest, most modern mechanical enterprise in the 1840s was a huge gun-making complex in southwestern Massachusetts. There, public and private innovation flourished together. Technique was reproduced at scale. This place of communion was called the Springfield Armory.

The Armory had been erected on an elevated plain near the confluence of three rivers a half mile from the village of Springfield, whose residents feared that unruly, drunken laborers would disrupt their bucolic community after work if they were too close to town. Springfield Armory was born from war, thanks in part to George Washington, who lamented the emerging country's dependence on foreign gunmakers. As commanding general, Washington pushed for the establishment of two domestic government armories, one in New England and the other in Harpers Ferry, Virginia. Springfield—the first deep-inland non-native settlement in the broad, long Connecticut River Valley—was chosen in part because it lay too far upriver to be attacked by the Royal Navy. That region would eventually be nicknamed Gun Valley.

Originally what would become the Springfield Armory was the Continental Arsenal, little more than a storage facility for gunpowder and various military supplies, where workers repaired small arms and made paper musket cartridges and vehicles to carry cannons. In 1794 legislation authorizing musket production transformed the Arsenal into the Armory. There, individual gunsmithing gave way to workers performing specific, limited duties. Private gunmakers and inventors came to take advantage of a public enterprise, where the sharing of ideas, including patents, was the order of the day, all for the common good of arming the nation. Skilled workmen in the region had at their core a Yankee inclination for inventiveness. Creativity abounded.

Springfield Armory brought the United States into a new era. It met the nation's growing needs by turning out guns with somewhat interchangeable parts. It was a product of a new America, where tourists flocked to watch mammoth machines make weapons at an astonishing pace. The spirit of co-operation between the federal government and private Yankee entrepreneurs yielded profound changes in what was manufactured and how. This vibrant productivity with mechanized production and interchangeable parts was tied to the United States to such an extent that the English dubbed it the American System of Manufactures, a system that would flourish in factories along the Connecticut River corridor, making Gun Valley home to the world's first machine-tool industries, the Silicon Valley of its time. Machinery and technology from there spread all over the globe. Because the transformation began with guns, the American System was also known as the Armory Practice.

Springfield Armory heralded a new age for the United States, in which the nation's growing military and industrial power was buoyed by the mass production of weapons. When the railroad came to Springfield in the early 1840s, it brought streams of tourists, who marveled at modern gunmaking in the largest metalworking establishment in the country. After visitors were shown how the manufacturing process worked, they gawked at displays of guns abundant enough to supply any army on earth. This was exactly the impression ordnance officer and engineer Major William Wade had in mind when he designed the armory's displays:

> The general arrangement of the interior pleases me much. It is something new, and I think well adapted to its destined purpose. The spectacle of a room containing twenty thousand arms, so

> arranged that every one would be visible; that any one could be taken hold of, examined, and replaced; at pleasure; with abundance of light, and of space for passages; the absence of any visible means by which they, or the floor above, are supported; the order, simplicity, neatness, and magnitude of the whole; would together, form a scene worth a journey of some miles to enjoy.

Partly because the machine age was in its infancy, the sight of massive, complex mechanisms inspired rhapsodies. "The whole scene appears more beautiful than warlike," wrote one visitor, "and it hardly seems possible, that an exhibition which fills the mind with such pleasurable emotions, can be made up of the instruments of death." Swooning over the armory, *The Springfield Republican* gushed, "The machinery here is absolutely poetical, both in structure and operation. It is pregnant with intelligence, rolls out its rhymes in beautiful measure, and sings of human ingenuity and the almost unlimited control of the human intellect over brute matter and the natural forces, with an eloquence which none but a clod of humanity can listen to without emotion."

Not every American felt at ease with Springfield's massive displays of weaponry—or the warlike impulses they reflected. In the summer of 1843, poet Henry Wadsworth Longfellow was on his honeymoon traveling through New England to visit his bride Fanny's relatives, when the couple decided to stop at the Armory along the way. Fanny was not only an elegant beauty and heiress, whom the Romantic poet had loved for years, but also a smart, artistic, and cultured aesthete who often suggested subjects for her husband to write about. Longfellow deeply respected her judgment. They both found a different kind of inspiration at the Springfield Armory.

In front of them were more than a hundred thousand new muskets made in the Armory, accompanied by scores made elsewhere, their muzzles pointed skyward in rigid formation through a series of rooms. Each barrel was cradled in an oiled walnut half-sleeve, like a uniform, that flared at the bottom. All were arrayed with exactitude, only an inch or so between them, in dozens of double-decked wooden frames painted gold. Forming tunnels in their lineups, tens of thousands of metal loops dubbed trigger guards gaped, ready for men's forefingers to summon the guns into service. But now the weapons, too new to have been blooded in battle, were silent.

Fanny, a pacifist, looked at the motionless parade of silent firepower and had an idea for her husband. Maybe, she urged, he could use the experience to write a peace poem. This place of nascent violence was the perfect inspiration. And so, Longfellow wrote:

This is the Arsenal. From floor to ceiling,
Like a huge organ, rise the burnished arms;
But from their silent pipes no anthem pealing
Startles the villages with strange alarms.

. . . . .

Were half the power, that fills the world with terror,
Were half the wealth bestowed on camps and courts,
Given to redeem the human mind from error,
There were no need of arsenals or forts:

The warrior's name would be a name abhorrèd!
And every nation, that should lift again
Its hand against a brother, on its forehead
Would wear forevermore the curse of Cain!

Down the dark future, through long generations,
The echoing sounds grow fainter and then cease;
And like a bell, with solemn, sweet vibrations,
I hear once more the voice of Christ say, "Peace!"

By the time Longfellow's ode for a peaceful world was published, two years after his visit to Springfield, American inventiveness—"Yankee ingenuity"—was in high gear far outside the Armory's walls. Entrepreneurs, smiths, speculators, dreamers, and machine-lovers throughout Gun Valley and beyond vied to make their marks and their fortunes, stoking the fires of America's Industrial Revolution as it gathered speed. The weapons these men developed would have far more power than the lethal force lying dormant in the rows of muskets that so awed the Longfellows and their fellow tourists. In the decades ahead there would be a terrible abundance of opportunities to use them.

Pivotal moments in the evolution of rapid firepower are many, Walter Hunt's inventions among them. So, too, are locations where firearms made the difference when it counted: not at orderly rifle ranges or peaceful test facilities or in Springfield Armory's vast machine shops, but on chaotic battlefields of the American West and the Civil War's many conflagrations. One key event took place alongside a slim flow of water two thousand miles to the south and west of Gun Valley on an early June day less than a year after the Longfellows visited Springfield.

1

DEVIL YACK

Yellow Wolf knew the Pinta Trail well, as had his ancestors and the Spanish and the Lipan Apaches and other tribes. Now at least some Anglos who had come to the Texas Hill Country knew it too. From a wooded hill above, he could see a group of about fifteen riders who had stopped where the trail crossed the Guadalupe River. The whites below were no match for his several dozen Comanches of the Penateka band. He would take them.

His warriors were armed, as they usually were, with lances and arrows, which they could launch repeatedly and with precision. Also, as usual, the whites had guns, fearsome weapons that roared and struck with power even at a distance, but each had to be painstakingly reloaded after a single shot. Comanches knew about guns; more than a few had them. They also knew that what the gun boasted in mightiness it lacked in versatility. At this moment in American history, a gun was not that great a threat when matched against flocks of incoming arrows.

What the Comanches could not know was that on this early June day in 1844, warfare on the plains would change forever. The Industrial Revolution had arrived in Texas in the form of a small piece of handheld weaponry born from the genius of Samuel Colt just a few years before. This gun would not exhaust itself after one shot. It would fire again in a second. And again.

And again. And again. Each of the fifteen white riders had at least one of Colt's inventions, probably two, tucked into his belt.

More than a century before, the Comanches had swept into Texas Hill Country from the north and the west, bringing their adaptable culture with them, on horses introduced onto the continent by imperial Spain. Now, thanks to these fleet and strong animals as well as their adaptive, resilient culture, Comanches were lords of the plains, having pushed aside the Spanish and other Indian nations to set up an empire of their own.

At first the vast domain called Comanchería had little problem with Anglos arriving from the East at the invitation of Spanish-speakers in now independent Mexico, for the fair-skinned newcomers brought trade that could enrich the Comanche empire's powerful reach. Now things were different. The new Republic of Texas granted land to an influx of settlers who took over territory the Comanches thought Texas had no right to claim. Surveyors cut the plains into parcels on maps that would define whose property was whose, ready to be transformed from "wilderness" to "civilization" by newcomers' willing hands.

For Yellow Wolf and his people, these surveyors and settlers had become the enemy. His and other Comanche raiding parties attacked white homesteads hard, killing many, taking captives, and generally making life on the Texas plains a risky undertaking for alien families intent on taking root in or near Comanchería. No Hill Country settler felt secure. All knew of homesteads raided by Comanches, who treated harshly many of those they encountered. Even though they may not have known her personally, Texians were familiar with the story of Matilda Lockhart.

In the autumn of 1838, thirteen-year-old Matilda and four children from the neighboring Putnam family had just finished gathering pecans in the bottomland near the Lockharts' homestead on the Guadalupe River when Comanche raiders grabbed them, lashed them with rawhide thongs to Indian horses, and whisked them away to the Guadalupe Mountains. Two rescue expeditions into Indian country ended in failure. More than a year would pass before Matilda was reunited with her family. By then she was unrecognizable.

Matilda arrived in San Antonio in March of 1840 with a delegation of Penateka chiefs and warriors interested in negotiating a treaty with the Texians. Setbacks suffered in their domain made the Comanches see this as a time to make peace. Cheyenne and Arapaho parties had threatened Comanchería's northern frontier, and Rangers had successfully harassed Comanches elsewhere, preferring to catch the Indians by surprise in their villages, as the Indians were doing to white settlements. And then there were cholera and smallpox, several recent epidemics having ripped through the Penateka community. The Comanches were ready for calmer relations with the Texians.

Return of captives was among the demands made by the Texians two months before, so Matilda Lockhart was with the Comanche peace delegation led by a chieftain named Muk-wah-ruh. A woman who helped bathe and dress the now-sixteen-year-old Matilda on her return found her "utterly degraded," a girl who "could never hold her head up again." It was not just what Matilda told them that enraged the Texians. It was what they saw.

"Her head, arms and face were full of bruises, and sores, and her nose actually burnt off to the bone—all the fleshy end gone, and a great scab formed on the end of the bone," wrote Mary Ann Maverick, who cared for the newly released captive. "Both nostrils were wide open and denuded of flesh. She told a piteous tale of how dreadfully the Indians had beaten her, and how they would wake her from sleep by sticking a chunk of fire to her flesh, especially to her nose, and how they would shout and laugh like fiends when she cried."

Matilda told the Texas commissioners negotiating with the Penatekas about a dozen or more additional white captives, whom the Comanches planned to offer up, one at a time, in exchange for various supplies. That was not part of the deal, the Texians told Muk-wah-ruh in the Council House, a one-story, flat-roofed stone building with an earthen floor, which was the usual place in San Antonio for serious talks between whites and

Indians; he was supposed to have brought all prisoners at once. This was impossible, Muk-wah-ruh explained. Those captives were held by other Comanche bands—not the Penateka—over whom he had no authority. The commissioners, enraged by Matilda's treatment and fearful that other children were being tortured, were not going to renegotiate. On their order soldiers entered the Council House to hold the Indian negotiators hostage until all white captives had been freed. The Comanches inside tried to escape, calling on tribesmen outside to help. Gunfighting erupted, killing most of the Indians inside the Council House, including Muk-wah-ruh. In the end the Texians seized more than two dozen Comanches, whom they offered to return once the white captives came safely home. Penateka leaders ignored the offer, and most of the Indians held by the Texians eventually managed to escape. There was no more talk of peace.

Broken by her ordeals, Matilda Lockhart never recovered. She died before she turned twenty. Texians would remember.

The Penateka would also remember. For them, the slaughter of peace ambassadors was unforgivable. They believed that the treacherous Texians' plan all along was to hold Muk-wah-ruh and his negotiators at gunpoint until every white captive was freed. The ones who had done this evil and those who benefitted from it would pay. There would be no suspension of raids into Texas—another of the demands Texians made at the beginning of the failed peace negotiations. Instead violence would escalate, as Comanchería clashed with the new Anglo empire taking hold in what the Indians considered their domain.

Wildflowers and greenery roused from a Hill Country winter had given way to early summer, June warming the tough terrain eighty miles northwest of San Antonio, a land of limestone and clear streams over which many battles had already been fought. The white men Yellow Wolf watched on the Pinta Trail by the Guadalupe River wore no uniforms, carried no flags, but the Penateka veteran of many skirmishes recognized them as Texas Rangers, a loosely bound collection of stalwarts deserving of Comanche respect, something the Indians did not give freely. The respect was for the Rangers' hardiness in combat, not for their role in protecting or avenging those the Comanches saw as intruders.

Despite the Rangers' fighting abilities and horsemanship skill that rivaled the Comanches', Yellow Wolf knew that numbers gave him the edge. He and a handful of his men would make their presence known. The rest would lie in wait above the river, obscured from view by live oaks, hardy trees that held most of their leaves through the harshest of prairie winters and were now bursting with foliage perfect for concealment. Tricked into thinking they faced only a few Comanches, the Rangers would go after Yellow Wolf, only to be slaughtered when they reached the oaks. It was a common tactic; if played right, it would work.

The fighting on the Pinta Trail would likely be fierce. The Texas Rangers knew that Comanches gave no quarter. But neither did the Rangers, who understood that surrender was never an option for them, because it only meant death often preceded by something worse.

Captain John Coffee "Jack" Hays and the fourteen Rangers Yellow Wolf was watching had left San Antonio a week before to scout for Indian bands—possibly Mexicans, too—who had been raiding white settlements. Now, after no Indians had been found, they headed back.

A slender, five-foot-eight, naturally pale fellow with a boyishly smooth face and gaunt cheeks weathered by the frontier, Hays did not look like a man to lead Texas Rangers—and lead was the right word, for they couldn't be commanded. Nor did he sound the part, with a quiet voice fitting his gentlemanly Tennessee upbringing. He tended not to talk much anyway. His clothing style was as modest as his demeanor—often a black leather cap with a blue roundabout jacket and black trousers—yet another contrast to the brawny Rangers and their broad-brimmed hats, which protected them from the Texas sun. It was said that his restless hazel eyes often looked sad. On foot Captain Jack walked slightly stooped, a tendency some thought made him look nervous, though losing nerve was not among his traits.

Orphaned at fifteen, Hays had headed west to Texas four years later in 1836, the year the Alamo fell and Texians rose in righteous fervor to wrest their independence from Mexico. He worked first as a surveyor and soon joined the Rangers, where he rose quickly to the rank of captain. By then Hays had proven his ability as well as any man to withstand the terror of battle and the hardship of long sojourns over the plains and show no strain for his trouble. At one with the terrain, Hays could divine the presence of passing Indians from the tiny pebbles displaced by their horses, even reading there the direction in which they rode. According to a Ranger who had served with him from the early days, "no officer ever possessed more completely the esteem, the confidence, and the love of his men."

Indians agreed that Hays was a man of substance. The Lipan Apaches, who were no friends of the Comanches and often allied themselves with the Rangers, called him bravo-too-much. "Me and Red Wing not afraid to go to hell together," explained Lipan chief Flacco, who sometimes fought alongside Hays. "Captain Jack heap brave, not afraid to go to hell by himself." The Comanches also had a name for him: Devil Yack.

On their journey back to San Antonio, Jack Hays and his men crossed the Guadalupe River near a smaller flow of water later called Walker's Creek. When the Rangers saw a beehive hanging from a tree at the crossing, they decided to take advantage of their good fortune. Honey was a delicious luxury. Its sweetness would complement nicely the meager rations each man had brought with him and the venison the plains provided.

The day was still fresh, and so were the Texas Rangers. Rangers traveled light, as they had to, even though their sturdy, mixed-breed mounts could support weight over long rides. They tended to be big men and brought no more than they needed, so their horses could be agile in battle. The Rangers were armed, of course, mostly with pistols and long knives but also with rifles. But on this mission each man had a new weapon: a different kind of pistol, one with a nine-inch barrel leading to a revolving cylinder just above and in front of the grip. Aside from the barrel and the method of holding the weapon, this handgun—a ballet of rods, screws, plates, and a couple of curved projections flowing into an elegant handle of American walnut—was unlike any used before in combat.

Its inventor was the flamboyantly driven Samuel Colt, who claimed he got the idea as a teenager by watching a ship's wheel or windlass turn and be locked into place by a wooden stave. In Colt's imagination, so he said, the spaces between the spokes became five hollow chambers, each closed at one end. The stave became a ratchet holding the cylinder firm at precise intervals as it turned. Fill each chamber with gunpowder and a lead ball a bit over a third of an inch thick and line up the chambers one by one with a long tube as you cocked the hammer, and you had a gun whose trigger would pop out beneath the cylinder. This was a gun that could shoot five times without reloading, its cylinder revolving after each shot—provided it worked as intended. Colt's creation did tend to be finicky compared with the less complicated mechanism of a single-shot pistol. Sometimes the mechanism failed. Or all five chambers would go off at once in a conflagration that endangered the man who pulled the trigger.

Whether Colt's idea came to him aboard ship or, more likely, from seeing a flintlock pistol with a hand-turned cylinder containing chambers, the truth is elusive. Whatever the inspiration, Colt started making his "revolver" in 1837 at a factory in Paterson, New Jersey, a hub of early industrialization where he had family contacts, but he could not entice the United States government, his target market, into buying it in bulk. Less than a year after production began, he personally took ninety of his "Paterson" rifles to Florida to convince the Army that they would help defeat Indians in the Second Seminole War. Colt sold more than half his supply, but although they performed well enough, he got no serious contracts. His efforts included giving personal "inducements" to powerful individuals, including the Army's chief of ordnance in 1839. This led the Colt firm's treasurer—Sam's cousin and a major shareholder—to declare in writing, "I will not become a party to a negotiation with a public officer to allow him compensation for aid in securing a contract with Govet." Colt paid him no mind and continued to offer inducements. He would do anything for success.

Colt did persuade one government to buy three hundred sixty of his revolving carbines and handguns: the Republic of Texas. When the Texian Navy was decommissioned in 1843, a number of Colt's small industrial masterpieces came into the hands of the republic's Rangers, though never as standard equipment for an entire company until Jack Hays found them appealing. Now, on their June 1844 mission in Indian country, every Ranger with Hays on the Pinta Trail had at least one, plus a loaded second cylinder that could be swapped for an expended one even in battle, if the gunman had the calm fortitude to do the necessary tinkering while under fire. Reloading a spent cylinder, however, was a task that needed quiet time. And a revolver without a loaded cylinder was no better than a club.

Yellow Wolf could see the Rangers in the distance but not the pistols stuck in their belts. Even if he had been able to make them out, it wouldn't have made any difference. A gun was a gun: one shot and done. When it went off, it had to be reloaded, and that required time and attention, which a warrior would exploit. As the white men fiddled, the Comanches would launch their arrows. Yellow Wolf ordered a contingent to ride ahead and bait the white men into an ambush.

On Hays's orders, two Rangers had lagged behind to see if their group was being followed—a standard Indian practice Hays had adopted. While Ranger Noah Cherry was mining the beehive atop the tree, the pair came galloping back to camp to report that there were, indeed, Comanches on their way. Then Cherry saw them too.

"Jerusalem!" he shouted from his perch. "Captain, yonder comes a thousand Indians!"

Cherry exaggerated; there were only about ten. Hays ordered his men to mount. The Rangers turned toward the Indians, who pivoted for the nearby hill thick with oak. Hays, familiar with Comanche tactics, knew this was a trap.

When he saw the Rangers holding to a slow walk, Yellow Wolf and the entire seventy-five-man Penateka band emerged from the hilltop oaks. Neither retreating nor charging, Hays and his men continued moving slowly forward, in no apparent hurry to engage the warriors. Seeing no assault coming, the Indians taunted the Rangers with shouts of "Charge! Charge!" in both English and Spanish, with a few epithets thrown in.

The Rangers then wheeled and launched into full gallop—not in reverse, but splitting into two groups across a shallow ravine out of sight of the Indians. The groups took opposite sides of the hill, circling behind their enemy and, despite a five-to-one disadvantage in manpower, broke cover and charged into the Comanches' midst.

After receiving several bullets from the Rangers' single-shot rifles, the warriors counterattacked. But by that time, the highly disciplined Rangers had regathered, circled their horses rump to rump, and begun firing their five-shot Colts at the charging Comanches, as the Indians shot arrows and thrust lances on the gallop. Rangers Robert Addison "Ad" Gillespie, a Tennessean like Hays, and Samuel H. Walker, a transplanted Marylander a month younger than the captain, were lanced but kept fighting.

In mid-battle, the Rangers replaced their spent cylinders with loaded ones. Some men who had fired all their ten shots even maintained enough composure between Comanche charges to reload cylinders still warm from having fired themselves empty. Yellow Wolf's band now realized that the Rangers were more dangerous than before. One Penateka later complained that their white adversaries "had a shot for every finger on the hand."

The Comanches regrouped at a distance. They didn't panic, but they had to plan quickly. Twenty of their men lay dead. Others were wounded. While the cluster of Rangers appeared to be intact with few casualties, Yellow Wolf's warriors still outnumbered them. The Comanches would not yield, but they also could not gauge how many shots the Texians had left in their newfound weaponry.

Hays knew, or at least he could make a good guess: The Rangers' ammunition was about gone. No matter how valiantly, how cleverly they fought, they could not withstand another charge. The Comanches were arrayed beyond the effective reach of whatever loads remained in Colt's revolvers. But it was not too far for a long-barreled rifle.

"Any man who has a load, kill that chief!" shouted Hays. One predictable thing about the Comanche he knew: Take out the leader, and there was a good chance they'd be done for the day. At this point it was the Rangers' only chance. Yellow Wolf sat erect on his horse, a shield covering his chest.

"I'll do it," said Gillespie, who cradled a still-loaded rifle. He got down from his horse to make sure he had a steady hold on his weapon, for steadiness was required when shooting precisely at even modest distance. One shot was all Gillespie had, and he hoped it was all he needed. Despite the lance wound through his side, he aimed calmly down the barrel, nestling the front sight above the muzzle into the V-shaped notch close to his eye. The figure of the large mounted man he targeted was blurry, but the post was in focus; that's what mattered. Gillespie squeezed the trigger.

The Penateka were studying the bunched-up Rangers out in the open, weighing their next move, when Gillespie's bullet found its mark. Yellow Wolf reeled in his saddle before tumbling to the ground.

Seeing their chief fatally shot, the remaining Penatekas broke. Within minutes they were gone, leaving twenty-three dead warriors behind. Thirty Indians had been wounded. One Ranger, a German immigrant named Peter Fohr, had been mortally wounded by an arrow through his body; several other Rangers were crippled. Despite grave wounds Gillespie and Walker would recover. Though the chief was toppled by a single-shot rifle, it was Colt's five-round weapon that won the day for the Rangers. They were smaller in number, but better endowed—and faster—in ammunition.

The Paterson revolvers "did good execution," Hays wrote in a report to the Texas Secretary of War and Marine. "Had it not been for them, I [do not] doubt what the consequences would have been. I cannot recommend these arms too highly."

Samuel Colt could not have imagined a better endorsement of his revolver, but it came too late. His gunmaking venture in Paterson, New Jersey, had gone bankrupt two years before. Thanks to Colt's revolver, Samuel Walker survived the battle with Yellow Wolf's Comanches. Thanks to Walker, Samuel Colt would resurrect his gun business three years later.

2

RISE OF THE SHOWMAN

What excited Samuel Colt ever since he was a boy was getting attention and making things explode, often achieving both at the same moment. Now, in 1844, he had another chance to do just that. While Jack Hays's Rangers battled Comanches in Texas, Colt was back East, pitching his latest project: blowing up ships. With his revolver business defunct and the United States concerned about protecting its shoreline from potential invaders, profits from underwater mines seemed promising to the cash-strapped inventor.

"To all whom it may concern," he wrote to the US Patent Office on June 8, 1844, "be it known that I Samuel Colt . . . have invented a new and useful mode of using ammunition for Military purposes, and more particularly for the using of gunpowder to make submarine explosions in such manner as to destroy vessels when under sail in harbors or channels. . . ." The idea of deploying remotely controlled underwater explosives for harbor defense had been with Colt a long time, even before he launched his revolver company. In fact his attraction to exploding things floating peacefully on water went back at least to Independence Day two weeks before he turned fifteen.

"Sam'l Colt Will Blow a Raft Sky-High on Ware Pond, July 4, 1829"—so proclaimed a handbill Colt distributed throughout the Massachusetts community where his father, Christopher, ran a textile mill. With that, the boy's passion for showmanship burst forth. Like all good self-promoters,

he had put his name up front, so that spectators would remember who had dazzled them that day.

And dazzle them he would, Colt knew, because he had devoured the lessons in *Compendium of Knowledge,* a volume filled with scientific wisdom of the day, and had experimented with dyes and liquids and assorted materials in his father's mill. He had also tinkered with electricity and gunpowder and discovered a way to produce an explosion by sending electricity through wires beneath a pond's surface to a batch of gunpowder. To make the event spectacular, something had to blow up. That's where the raft would come in. When the hapless vessel was over the explosive, Colt would flip a switch from shore, and with much spray, thunder, and smoke, the raft would be gloriously sent aloft in splinters. This must be done in front of a crowd. Applause would follow, and the boy could take a bow. That was the plan. It would be grand!

July 4 was on a Saturday that year, which added to the festive nature of the holiday the hope that more people would be free to attend Samuel Colt's debut in the world of public explosions. Dressed in their Independence Day finery, ladies and gentlemen gathered at one end of the four-acre pond near the town of Ware to enjoy the show they had been promised. Colt, secreted in trees at the pond's edge away from the crowd, sent an electric charge speeding to the underwater mine. The surface erupted, the ground shook, and pond water doused the well-attired spectators. Colt, it seemed, had used too much gunpowder. To make matters worse, the raft had drifted away from the mine, leaving the eruption unchecked. The unamused crowd took after Colt, who managed to escape with the help of Elisha Root, a young mechanic who would later play a major role in Colt's business.

He was not done with pyrotechnics; the next July 4 Colt was back at it. While at Amherst Academy learning about navigation, the "young wild fellow," as a professor described him, joined a bunch of students in setting off a cannon on school grounds. According to the professor, this is what happened next:

> Some of the officers of College interfered & tried to stop the noise. Colt, as Prof. Fisk[e] ordered him not to fire again, and placing

> himself, as the story was told, the next day near the mouth of the gun, swung his match, & cried out "a gun for Prof. Fiske." & touched it off—The Prof. Enquired his name—& he replied, "his name was Colt, & he could Kick like Hell"—He soon left town, for good.

Expelled from Amherst, Colt signed on as a crewman aboard the merchant ship *Corvo,* sailing from Boston to Calcutta. It was during this voyage that Colt supposedly came up with his idea of a gun with a revolving cylinder that made it a multi-shot weapon. Whether the legend is fact or fiction, after returning from his eleven-month sea voyage, Colt embarked on a crusade to turn his idea into a moneymaking reality. The best way to scare up cash, Colt decided, was to blend his penchant for performance with his knowledge of chemicals. The perfect vehicle for that was nitrous oxide, otherwise known as laughing gas. Selling himself as the "celebrated Dr. Coult," he began touring the country, putting on exhibitions with willing patrons who breathed in the gas and delighted audiences with the results. He announced his presence in local newspapers:

> The public are respectfully informed that on THIS EVENING at half past eight o'clock, the EXHILERATING [*sic*] GAS, will be administered to visiters [*sic*]; these [*sic*] who are desirous of taking it will do well to apply early as the gas was exhausted at an early hour on the last evening of the exhibition. N. B. An evening will be set apart for ladies of which due notice will be given.

When inhaled, an advertisement in the *Albany Journal* told people, the gas produced "the most astonishing effects on the nervous system; that some individuals were disposed to laugh, sing and dance; others to recitation and declamation, and that the greater number had an irresistable [*sic*] propensity to muscular exertion, such as wrestling, boxing, &c. with innumerable fantastic feats." Because "Dr. C." was "a practiced Chemist," no one needed to fear inhaling something "impure." As for ladies interested in seeing the demonstrations, the advertisement assured "that the house enables every accommodation for their comfort, and that not a shadow of impropriety attends the exhibition."

Not every reaction to inhaling Colt's gas was harmless exhilaration. Sometimes, as this report from *The Pittsburg Manufacturer* suggests, the effect was just the opposite:

> During four evenings of this week, a gentleman called Dr. Coult, exhibited at the old Museum the effects of Nitrous Oxide Gas on the human system. The scenes that ensued from inhaling it during the exhibition, beggars all description. Some danced and jumped, others cut up singularly fantastic tricks, but the greater portion of those who inhaled it became extremely pugnacious. Some of these, though placed within a strong enclosure, managed to escape from it, and attack and beat unmercifully the audience. One strong fellow, who became on taking it as furious as an enraged lion, sprang over the enclosure, and drove every soul out of the room into the street, beating two or three very severely.

Colt found receptive audiences for his laughing-gas shows up and down the East Coast, while he perfected his hawking skills and harvested cash. Ever the glad-hander, he had become friends with the director of an ersatz museum in Baltimore, who gave him access to its lecture hall for his shows. Colt decided to stay there for a while, adopting the town as his base of operations. Perhaps he could find a local gunsmith skilled enough to make patent models for his revolver. Like most towns, Baltimore had its share of craftsmen who made things that shot, including rifles, pistols, and "every article for the Sportsman's use," as gunsmith Arthur T. Baxter promised would be available at his shop near a busy downtown wharf. "Captains of vessels bringing work, are requested to leave it as soon as they arrive, so as to give as much time as possible. This will prevent disappointment."

Baxter was a good bet. He was doing well enough to hire several assistants to handle the workload. One of those gunsmiths, English-born John

Pearson, who had started his career as a watchmaker, was assigned to work with Colt and was eventually left alone to make whatever independent arrangements he wanted. For Pearson this was a mistake, despite the contract the two had signed, which guaranteed that Colt would pay him ten dollars each week for a year.

Colt had a habit of being loose with money, especially when it belonged to others, which continually put Pearson in a squeeze. The inventor would promise to pay the bills and sometimes did, but not enough to keep the hapless gunsmith solvent. "I am out of money," Pearson wrote to Colt, "and the rent is due today and I want some more wood for fire so you must send some money immediately or I shall be lost." Eventually Pearson was paid. Colt got his patent models and left Baltimore for good.

When the company making his first revolvers—rifles and handguns—went bankrupt, Colt still held the patent and dreams of returning to the gun business. There had been other guns with revolving cylinders, but they required manual turning in order to line up chambers with the barrel. Colt's forward step was revolving the cylinder while the gun was being cocked and then locking it in place for shooting. The idea was sound, he thought, though it needed work. In the meantime, he had other ideas that, of course, involved explosives. He invented waterproof tinfoil cartridges for guns, which were making him some money. And then there was his idea for protecting the East Coast with underwater mines in case of invasion.

Exploding subsurface mines to sink ships was not a new idea. Steamboat pioneer Robert Fulton had worked on them. So had others, to the dismay of some, including British politician and military figure Sir Howard Douglass, who called mines an "inhuman system prepared for naval warfare in an age of enlightened humanity, . . . a merciless, barbarous idea." Now it was Samuel Colt's turn, despite the objections of congressman and former president John Quincy Adams, who agreed with Douglass and lambasted the use of mines as "cowardly, and no fair or honest warfare." Colt's contribution was a way to blow up a mine at the moment a ship passed over it. How he did that he kept a secret from everyone except people who could get him money and political support. Colt eventually got enough of both to demonstrate what he could do with a steamer chosen for the occasion off Washington in late August 1842 before an eager throng that included a host of dignitaries, including the president of the United States. When Colt detonated the device from shore, "a magnificent and astonishing spectacle was presented to us," exulted a journalist on the scene.

> The water around the vessel was upheaved, and rose in a vast and majestic column, to an astonishing height—a gigantic jet d'core—a marine volcano. No comparison can give an adequate idea of its grandeur. As to the vessel, she was not visible in the mass of foam and water; but the thousands of small, dark splinters into which she was shredded, were seen raising with the upper mass of the column, into the air.

Colt blew up more ships, including a schooner in New York on a lovely October day. After that ship was destroyed, souvenir hunters in boats floating nearby dashed in to pick up fragments. And on the Potomac River near the Washington Navy Yard in April 1844, Colt destroyed a five-hundred-ton three-masted ship under full sail, the stars and stripes waving from its masthead, a vessel "doomed to be offered up [as] a sacrifice for the improvement of science and the extension of human knowledge." The House of Representatives had adjourned early that day to enjoy the spectacle.

But Colt was not making headway in getting the contracts he wanted. For one thing, his mania for secrecy undercut him. He also overspent money the government had given him. Making his case even more difficult was that a number of people he had enlisted on his side were either no longer players or were dead. He kept pushing anyway. If love of country drove him forward, there's little evidence of it. What propelled Colt was a desire for success. His Hartford-based family had had financial ups and downs, and with those fluctuations, respect in society for Colt and his close kin also shifted up and down beyond his control. Colt yearned not for heroism in battle, though sometimes he claimed he did, but for riches and respect wherever he could get them. In the early 1840s he certainly received lots of attention, but as 1844 ended, his efforts to convince the government to buy his underwater mine system continued to lose ground. Eventually interest in what Colt had to sell evaporated.

Interest in another Colt product, however, would be rekindled. As 1844 wound down, the expansionist mood in the country would find its champion in a new president and the mounting prospect of war, this time with Mexico. Word of the Texas Rangers' success with Colt's revolver against Comanches was spreading. One Ranger would soon reach out to Colt with ideas for how to make the gun better. Colt's fortunes were set to rise.

3

MASTER OF STEEL

According to the tale of the Remingtons that has come down through the decades, fact and fantasy, company founder Eliphalet Remington II had a poet's heart. When news arrived that the War of 1812 had ended—meaning that his family's home in New York State's Mohawk Valley was now safe from British invasion—the elated twenty-one-year-old Remington burst into verse:

Hale sacred peace, thy gentle reign
Is now restored to us again,
Thy radiant smiles and gentle voice,
Bid every virtuous heart rejoice.

But can thy smiles disperse the gloom
That reigns within the warrior's tomb,
Can it assuage the widow's grief
Or to the orphan speak relief?

Despite poetical flourishes, the Remingtons were serious, solid folk—some might call them dour—not much given to flights of fancy or casual humor, though "Lite," as he was called, had been a romantic youth with pacifist leanings. His mother, Elizabeth, saw great potential in Lite's daydreaming and versifying; perhaps he could be a celebrated author. What others may have considered a frivolous waste of time, she encouraged. What farmer, carpenter, toolmaker, blacksmith, agricultural equipment–forger Eliphalet I thought of his son's literary potential can only be guessed at. Work and devotion to God's plan on earth, as well as a thirst for success, compelled the Remingtons onward. A love of the written word also ran through the family; Lite devoured John Milton's *Paradise Lost* at least three times and never stopped writing poetry. Although his lifelong passion, if not his skill, for rhyme and cadence was strong, Lite's true talent would always be in metal.

Eliphalet I and Elizabeth Remington with their son and three daughters had come to the Mohawk River Valley from Suffield, Connecticut around 1800. Both parents traced their roots to Puritan ancestors who had left to cross the Atlantic in the Great Migration of the 1630s, settling in Massachusetts and Connecticut, where the families stayed for the next century and a half. The Remingtons, along with other Connecticut families, decided that a move northwest to the newly surveyed but sparsely settled wooded hills of central New York would improve their prospects. Eliphalet had been doing well as a part-time farmer, but the rocky New England soil was always a challenge. So, as the eighteenth century wound down, he, Elizabeth, and their children (Lite was seven) piled into oxcarts for the rough overland ride toward new country. Their trek ended about 170 hard miles later in high, rolling terrain at a tiny village called Litchfield in Herkimer County. They were now on the frontier.

The Remingtons were hardly poor when they arrived in New York State. Eliphalet was able to buy a fifty-acre plot of land, a holding that he eventually expanded to three hundred acres. It was not all usable for farming, but that was fine with Eliphalet. He wanted to take advantage of a creek flowing swiftly northward at the bottom of a steep, rock-walled gorge, on its way to join the Mohawk River several miles away. The creek's propulsive flow was a perfect power source for Eliphalet's business plans. Ten years after the Remingtons arrived in New York, Eliphalet built a forge from fieldstone, where he turned cast or pig iron into wrought iron to fashion a variety of tools for the growing community, including plows, crowbars, and axes. The business was a success, the forge running steadily, closing only when the water that drove its machinery froze during the bitterest stretches of winter.

If 1814 was a good year for the father, it was an even better one for the son. On May 12, Lite married Abigail Paddock, another Connecticut emigrant nearly three years his senior who lived with her parents a quarter mile away from the solid, two-story stone house Eliphalet had built for his family. Abigail was a relative latecomer to the area, her parents having remained in Connecticut during her early years. That meant she received more schooling than Lite did, since formal teaching was rare on the frontier. No matter. Lite had always been drawn to books, which made him both self-educated and interesting to her. He also had all the makings of a good provider. The young couple lived in the elder Remington's home until two years after they started a family of their own, which they did in 1816. They named the first of their five children Philo. That same year, Lite started another kind of family when he made his first rifle barrel. The Remington firearms company was born.

The origin story goes like this: Lite wanted a rifle, but his father refused to give him money to buy one.

"Eliphalet Jr. closed his firm jaws tightly, and began collecting scrap iron on his own account," according to one old history. "This he welded skilfully into a gun-barrel, walked fifteen miles to Utica to have it rifled, and finally had a weapon of which he might well be proud." The gun was so good, according to legend, that the people of Litchfield loved it. Such workmanship! Such accuracy! "[S]oon the neighbors ordered others like it, and before long the Remington forge found itself hard at work to meet the increasing demand. Several times each week the stalwart young manufacturer packed a load of gun-barrels upon his back, and tramped all the way to Utica where a gunsmith rifled and finished them."

It's a nice story, but it's largely untrue. The company itself promoted this frontier legend, which hinged on the romantic ideals of individual achievement. The year was right—1816—and Lite *did* make a barrel. But what he is better credited with doing after that was not shepherding guns into the hands of eager neighbors but coming up with a way to make barrels less expensively and, more important, stronger.

Remington barrels were making their mark. Increasingly gunsmiths ordered them for their accuracy and quality. While the little factory in the gorge could handle orders for the variety of implements the Remingtons had been turning out, the barrel business threatened to outgrow the place. The land between the looming rock walls on either side was too narrow for the forge to grow. The Remingtons had to find a new location, and the timing could not have been better. Just twenty-eight miles away New York State was about to break ground for the Erie Canal, an engineering marvel that would open up the frontier, boost the Remingtons' fortunes, and change the country.

On Independence Day 1817, the year after Lite made his first gun barrel, a spadeful of earth was turned just outside Rome, New York, northwest of the Remington forge. The moment was celebrated with grand ceremonies attended by dignitaries, including newly elected New York governor DeWitt Clinton, who reveled in the occasion he had so longed to see. Digging for the Erie Canal had begun.

Canals had been dug elsewhere, especially in industrializing Europe, to speed goods to market inexpensively, so the idea wasn't new. It wasn't new in the United States, either. George Washington, for one, had urged the creation of a great canal reaching westward to the Ohio Valley, but he envisioned it beginning near his Virginia property. The Erie route between Buffalo at the Great Lakes and the Hudson River near Albany made sense, because it would link the interior to the Atlantic Coast and, of course, to New York City. President James Madison vetoed a bill that would have provided federal funds for the canal, but Clinton persuaded his state to support the project. There were fears of a north-south "dismemberment of the Union," Clinton said, and while they were genuine, a more pressing danger was the threat of a division between states on the coast and those in the interior. A grand canal linking East to West "will form an imperishable cement of connection, and an indissoluble bond of union."

Clinton believed the canal would benefit the entire country, not just his home state. Others agreed. No longer would inland producers—farmers and metal makers alike—be forced to lug their goods at great expense over rough roads like the ones traveled by the Remingtons on their way to New York. More markets, and bigger ones, would be within reach. While the challenge was enormous and the geological obstacles daunting, engineers did have one advantage in cutting through the landscape: A Delaware chemical company, E. I. Du Pont de Nemours, had come up with a new kind of blasting powder with much more explosive force than the black gunpowder used to date.

The formal opening of the Erie Canal in 1825—all 363 miles of it, 40 feet wide and four feet deep, with 83 locks to ease boats down from the hills to the Hudson—was attended by even greater celebration than its groundbreaking eight years earlier. A hope of rising prosperity filled everyone, including Eliphalet I and Lite Remington.

On New Year's Day 1828 Eliphalet bought a hundred acres of land fronting the Erie Canal and built a new forge. The canal was not quite three years old, and the settlement where Remington put down roots amounted to only seven homes, two storehouses, and a school. He built a low, single-story building so close to the man-made waterway that one side of its foundation formed a berm against the canal wall. Along the other side ran Main Street. The Remingtons were positioned perfectly to take advantage of the canal's potential. And freed from the cramped confines of the gorge, they had plenty of room to expand their business.

Lite had been gradually taking over running the company. His father still tended to making agricultural implements, but the barrel business was Lite's. As it grew, so did the younger Remington's role. Lite's increased duties required him to spend more time at the canal-side building, so he and his father decided that a structure should be raised nearby to accommodate his needs. Building was among Eliphalet's skills, so he took on the task.

Lumber for the new building came in twenty-foot planks from large trees that grew near the gorge. These Eliphalet I loaded onto drays—flatbed wagons without sides used for heavy loads. Thick straps and wooden holding stakes kept the lumber from falling off on its way to the building site. On the morning of June 22, 1828, five months after the move to the canal began, Eliphalet I climbed atop a load of lumber, while a young employee grabbed the reins of the four-horse wagon team. Along the way they started down a steep hill, and the driver leaned hard on the long brake to keep the wagon under control. Then they hit a sinkhole, and the load lurched, throwing Eliphalet I forward and under the rolling wagon. The driver pulled the reins as hard as he could and jammed his foot on the brake, but the wagon kept moving, one of its metal-rimmed wheels rolling over Eliphalet I. Remington was taken back to the stone house he had built in the gorge, the one in which he had raised his family and cultivated his business, and there, five days later, he died of his injuries.

Lite now shed his "II." He had become the senior Eliphalet, the man in charge of the family and the business. Late that autumn, on November 12, there would be another Eliphalet, when Abigail gave birth to their fifth child.

Despite his lifelong devotion to poetry, Lite Remington was a serious man intensely interested in business. Through the years he increasingly looked and acted the part. He dressed well and had taken to wearing a top hat in which he carried his business papers. Remington's physique complemented his attire. His "long, loose-jointed body" was topped by a "long and well-shaped [and delicate head], with curly, dark hair growing to a widow's peak on the high forehead. Lips and nose were finely chiseled, eyebrows delicately sketched above large, dreamy eyes." His tall, muscular build was "capable of great endurance. His manners were gentle and kindly, but his resolutions were firm, and obedience was enforced in the execution of his plans."

Unlike many barrel makers, Remington had a large plant that could handle a major government contract. Then he built up his factory even more. In 1837 eldest son Philo joined his father to make what was now E. Remington & Son a family business again. On an August afternoon four years later, as an old company history tells it, Abigail and her daughter Maria decided to take a drive back to the old Remington homestead.

After fording a creek leading to the stone house, Maria opened her parasol against the summer sun. She did it so briskly that the sound of the silk snapping open cracked like a gunshot. Their horse bolted, ripping the reins from Abigail's hands, and tore off, eventually sending the carriage into an oak tree. Amid the splintered wreckage, Abigail lay dead, her head crushed. According to his grandchildren, Remington never smiled again.

By the mid-1840s, E. Remington & Son was providing employment for much of the area. Most rifle barrel makers still fashioned their products by handheld hammers. Remington, however, set up machinery in his forge to do the job using power-driven hammers. This process cut the labor and time required to make a barrel, which enabled Remington to outpace his competitors. Remington was also making his barrels out of cast steel, not traditional iron. The forge's position right next to the Erie Canal helped speed his barrels to the smiths who would turn them into guns. Remington found a simple and quick means to get orders on their way to out-of-town customers. Through a hole in the floor of a canal bridge he dropped bundles of gun barrels onto line boats passing just below. This was fine, as long as orders were relatively small and spaced out, but he knew it would not do in the face of rising demand.

Remington's local prominence grew alongside both his business and the town itself. In 1844 mail was still delivered from the town of Mohawk, which was only a couple of miles away, but the growing community thought it was time for their own post office. After all, legend has it, the Mohawk postmaster brought their mail wrapped in a handkerchief, which meant he could deliver only what he was able to carry in one hand.

Some believed a new post office should be called Remington, to honor the man who had provided such abundant employment. Lite Remington disapproved, from modesty. It was one thing to have a business bear one's name; it was quite another to have it carry a town. In any event, according to old accounts, Remington showed his displeasure by refusing to use the Remington address, choosing instead another post office nine miles away. This meant that responses to his correspondence sometimes ended up elsewhere. Compounding the problem was the misdirection of Remington-addressed letters to several other places, including Bennington, Vermont, and a similarly named Pennsylvania town. Add to this the inevitable batch of illegibly written addresses, and the local mail lapsed into chaos.

Thought was put into finding a suitable Indian name for the town, but nothing satisfactory came to mind. Then a solution was offered. The local postmaster, David D. Devoe, suggested to his friend Remington that the town be called Ilion. (Devoe had been impressed when he read Homer's *Iliad.*) Remington thought that naming the little village after a classic site of such importance smacked of vanity. But in the end, he relented. Remington and Devoe then sent a petition to Washington asking that the name become Ilion, which it duly was.

Gaining a post office was a pittance in light of the national deal that now loomed for Remington. That same year Remington would take a big step toward becoming a mass producer of guns for the military. The United States government, already churning out weapons at the Springfield and Harpers Ferry armories, was eager to arm its soldiers by contracting with private firms. The government also wanted guns that could be shipped easily and quickly to the new St. Louis Armory for use in a region where more Americans wanted to go. Remington had gotten word that a particular gunsmith in Cincinnati was struggling to fulfill a contract with the government. English-born John Griffiths thought he could take advantage of the westward movement by expanding his usual business of making sporting firearms with a big—and highly profitable—military contract, so he had made a deal with the US Army Ordnance Department to produce five thousand state-of-the-art Model 1841 percussion rifles. By 1844 two years had passed since Griffiths had signed his contract with the government, and he had run into difficulties. His promises were more than his shop could deliver, and he needed rescuing. Word of Griffiths's dilemma reached Remington. Perhaps if he took over Griffiths's contract, he could adapt his operation as the Cincinnati gunsmith could not. It would be a risk, but the rewards could be great.

Without hesitation Remington headed for Cincinnati, where he found

4

MOBTOWN

Politicians gathering in Baltimore to nominate candidates for the presidency in May 1844 could have done worse than buy shirts from Oliver Fisher Winchester. Or other articles of clothing befitting the gentleman of the day or the lesser man eager to appear better. All were available, wholesale and retail, at the Winchester & Co. emporium downtown, near Baltimore's busy harbor. And with two national conventions coming to town with parades and festivities and gatherings of hard bargainers, it was the right time to outfit the expected influx of so many potential customers. "The city pulsated with excitement," wrote one historian. "As each new delegation arrived in town, banners appeared, music sounded, parades formed. All the hotels and boardinghouses were filled to capacity. Even private homes were thrown open to accommodate the swelling numbers of politicians and their followers."

While selling clothes may not seem a glamorous occupation for a man who would become a titan in the firearms industry, Oliver Winchester had already come a long way.

He and his twin brother, Samuel, were born in Boston on the last day of November 1810, the tenth and eleventh children of their father, also named Samuel, and his third wife, Hannah. The Winchester family traced its roots back to Puritans with a smattering of clerics along the line. In a privately published book in 1912, family historian Fanny Winchester Hotchkiss wrote:

> I do not find that they were men of note or known for their great deeds to mankind, but I do realize that they were men of strong character, earnest purpose and deep religious convictions, upright and useful citizens, holding many offices of trust in the communities of which they were members, and helpful in the upbuilding of the towns they selected for their homes, showing both ability and public spirit.

These Winchesters, she said, were "truly the pioneers of the new world."

Oliver and Samuel's father had not been as prosperous or as modestly noteworthy as many of his forebears, barely earning a living as a farmer. He left little in his wake except a scattering of children, the eldest a generation older than the youngest. He also left no impression on the twins, having died less than five months after they were born. The loss of the elder Samuel sent an already struggling family into poverty. The boys' mother did remarry, but the twins' new stepfather was not inclined to raise someone else's children. When the brothers were old enough to perform chores, they were put to work on a farm. All Winchesters were expected to work hard. Later Oliver would recall his childhood as "always hungry and always cold."

Oliver started school at the age of eleven, studying in the winter and farming the rest of the year to help the family survive. A carpenter opened up opportunities when he took on Oliver as a fourteen-year-old apprentice. The boy was now learning a trade, not a cerebral calling like the ministry of some of his ancestors, but an occupation that was worthwhile and portable. Boston was growing at that time and could use men who made things, especially for housing. But a stroke of luck before he turned nineteen lured Oliver down the coast to Baltimore, another town on the rise.

His luck involved another man's death. A much older half brother named William had long lived in Baltimore, where he had become moderately successful. When William died childless in 1829 at the age of forty-two, he left behind a respectable estate—including two lots of land in town that he had set aside in trust for his widow and several relatives, including Oliver and his three full siblings. His widow quickly went to court to have the properties sold and the proceeds divided among the beneficiaries. No one objected, including Oliver and twin Samuel, who were represented in court by their sister's husband, Cyrus Brett, since they had not yet turned twenty-one. Oliver now had some financial security that gave him a chance to begin a new life.

Baltimore and Boston were four hundred miles apart, a long distance for a teenager to travel solo in the early nineteenth century. But Oliver sensed brighter prospects down south, where he would soon have a financial stake of his own, away from a stepfather who cared little for him. He was just about the same age as his Puritan ancestor, John Winchester, two centuries before, when he abandoned England on the ship *Elizabeth* bound for Massachusetts. Setting a pattern that would continue through their lives, Oliver was the first of the twins to seek a new opportunity, this time heading south to Baltimore. Samuel would soon follow.

> *There is not to be found, perhaps, in the history of any country, certainly not in that of the United States, an instance of such rapidity of growth and improvement, as has been manifested in the city of Baltimore, during the last thirty years.*
>
> HEZEKIAH NILES
> SEPTEMBER 19, 1812

When editor and publisher Niles wrote that paean to his adopted hometown (he was a Pennsylvania native), some forty-seven thousand other people also called Baltimore home. By the time the teenaged Winchester twins arrived, Baltimore's population had reached eighty-one thousand. A decade later it would top a hundred thousand, making it the second largest city in the United States.

Early nineteenth-century Baltimore was a magnet for willing workers and for owners of slaves. Young craftsmen and manufacturers heard that they could find fresh opportunities in the city by the Chesapeake Bay, so they journeyed south from New England. Soon immigrants from Germany and Ireland would also seek their fortunes in Baltimore, first as laborers striving to climb higher. Despite a reputation for bawdiness, the town was beginning to flourish. A British visitor at the time noted that its shopkeepers, tradesmen, and mechanics were more informed, more industrious, and living better than their counterparts in England. Baltimore, a mere hamlet just a half century earlier, had rapidly turned into a metropolis, its perfect harbor having birthed a center of trade—including slaves—to and from Europe and the West Indies. The harbor and downtown wharfs bristled with ships' masts.

In the years before the Winchester twins arrived, local manufacturing had taken off. Exports and imports remained strong. On Independence Day the year before Oliver came to town, ninety-year-old Charles Carroll, last surviving signer of the Declaration of Independence, stuck a shovel into the ground to mark the beginning of the Baltimore and Ohio Railroad, another means of moving goods and people to benefit Baltimore, investors, and the nation. Civic pride abounded.

Once in Baltimore, Oliver Winchester sought the work he had been trained to do: carpentry. Of the twins it was Oliver who would one day rise to prominence. His skill—perhaps his drive, too—brought him employment, and he eventually became a master builder. Among his projects was erecting a new church. Before long Winchester gave up carpentry to go into the clothing business. First he sold shirts as an employee, but after marrying Maine-born Jane Ellen Hope, he decided to open his own store in which he sold a variety of men's furnishings close to Baltimore's retail and commercial section.

"Suspenders, Gloves, Hosiery, and READY MADE LINEN" were what Oliver Winchester offered customers, both wholesale and retail. There were other clothing makers in town, of course, some claiming the ability to outfit men in custom attire at great speed, a tempting option in a port city patronized by transients, many of them well-heeled. "Travelling gentlemen can have a suit of Clothes made at *six hours notice*," promised a tailor whose shop shared the same block as Winchester's.

While clothiers touted their ability to dress the well-off, their market was really more general than that. The first half of the nineteenth century saw an American democratization in attire. An Englishman's apparel revealed his social standing. This was less so for Americans, leading a British consul in 1840s Boston to lament the sight of servant girls "strongly infected with the national bad taste for being over-dressed," further describing them as "scarcely to be distinguished from their employers." Such displays would not have bothered Oliver Winchester; profiting from the mass market would always be his ambition.

Baltimore was not always a tranquil town. It was often seen as a disorderly, raucous place with periodic riots, including those that erupted in the summer of 1812 after the United States declared war on Great Britain and mere weeks before Niles lauded the city in print. Episodic mobbism had been a minor Anglo-American tradition, usually limited to property destruction, but "the speed and frequency with which the citizenry found excuse to riot" had earned Baltimore special recognition in the nickname Mobtown. "In the early days political feeling ran high and politics often was at the bottom of the trouble. However, when the populace was in the mood for going on a rampage almost any reason would do." Some saw rioting as a way to express grievances of the moment. The riots of 1812 were different. People died, and for a while that summer, disorder ruled.

True to its nickname, Baltimore erupted again around the time Winchester opened his first clothing shop close to the retail and commerce section of the city. Mobs rioted in August of 1835, enraged at directors of the Bank of Maryland, which had failed the year before, leaving many citizens penniless. Those who had lost money hoped for a financial settlement, but after months passing with no relief, violence broke out. Mobs destroyed the homes of two bank directors, throwing their contents into the street to be burned or, in the case of one man's collection of fine wines, carried away or drunk on the spot. People were beaten, and efforts by the mayor and the militia to quell the rioting were largely ineffectual. But the city survived, as did Oliver Winchester's business, despite the rioting and the financial crisis that hit the entire nation in 1837. By 1840 Oliver was doing well, now the father of two children: a daughter, Ann Rebecca, and a son, William Wirt.

He and Jane named their son after William Wirt, the US Attorney General and author who had prosecuted former vice president Aaron Burr for treason, run for president against Andrew Jackson, and was living in Baltimore when Oliver and Samuel came to town. Attorney General Wirt died three years before his Winchester namesake was born. Exactly what inspired Jane and Oliver Winchester to name their first son after this politician, prominent and respected though he was, is unclear. Some speculate that Oliver's character contained a touch of Wirt's philosophy: "Seize the moment of excited curiosity on any subject, to solve your doubts; for if you let it pass, the desire may never return, and you may remain in ignorance." Oliver did show an interest in politics, particularly the Whig Party. In 1840 Whigs had gathered in Baltimore, the "go-to city for US presidential nominating conventions," and won his support. Anyone interested in siding with anti-Jacksonian William Henry Harrison for president then could pick up campaign flags, Harrison handkerchiefs, log cabin buttons, and "hard cider stocks" (Harrison was the "log cabin and hard cider" candidate) at Winchester's store. Now, in 1844, the Whigs were back in town from all around the country, fervently boosting their current candidate, Henry Clay of Kentucky. "The whole place resembled a fair, and every street was alive with people, hurrying to and fro, chattering, singing, laughing. Music sounded and banners waved, as Whigs by the thousands poured into town."

"Clay badges hung conspicuously from every buttonhole. Clay portraits, Clay ribbons, Clay hats, Clay cigars, Clay banners, Clay songs, Clay marches, Clay quicksteps, Clay caricatures enveloped the city. 'Oh, the rushing, the driving, the noise, the excitement!' To see it and hear it and feel it was sheer ecstacy,' said one man."

While Oliver was peddling clothes, Samuel worked as a carpenter, though he would join his brother in the men's furnishings business. The Winchester twins were close, Samuel giving his second son his brother's name. With the 1840s well under way, Oliver's business success allowed him to offer financial help to other members of his family, including Samuel and their sister, Mary Ann Winchester Brett, and her many children. The brothers had children of their own; earning more money was becoming crucial. What if he could find a way to make clothing more quickly? What if he could make a better shirt, one that he could patent and would then be marketable as his exclusive product far beyond Baltimore? Oliver Winchester began experimenting.

5

FROM BAYONETS TO GUNS

Throughout 1844 Horace Smith busied himself with what would someday make him rich: building guns. The chance he'd become wealthy where he worked at that time was nil, for Smith was a mere employee helping to turn out firearms he had not designed and which did not bear his name. For several months the thirty-five-year-old Smith had been working alongside a dozen or so other men for Ethan Allen (not the Revolutionary War hero). Allen had a company, Allen & Thurber, which made a variety of firearms in Norwich, Connecticut, including a gun concealed in a gentleman's cane and a slim single-shot pistol dubbed the "pocket rifle" that could be discreetly tucked into a boot.

The most famous Allen & Thurber weapon, though, was a many-barreled pistol eventually called a pepperbox, since it resembled a household pepper grinder. This clumsy, front-heavy handgun boasted the ability to shoot multiple times quickly without reloading. It used a cluster of individual barrels that rotated with the pull of its trigger, giving the shooter wielding it several chances to send a lead ball where he wanted it to go. That multi-shot capacity was Ethan Allen's selling point.

That was also Samuel Colt's boast for his Paterson revolver, and Colt had been one year ahead of Allen in securing a patent. Their two pistols were different in design and usefulness, so there could be no patent battles. The only competition was in the marketplace. From Allen's perspective, however, there was no competition at all, as Colt's revolver business had gone under two years earlier.

Allen and his brother-in-law and moneyman Charles Thurber—the Thurber of Allen & Thurber—found a healthy market for their pepperboxes. So did other makers of this kind of gun, though Allen & Thurber dominated and had a patent for a particular model. A pepperbox mechanism was more complex than that of a standard pistol of the time, so the men who built them had to be particularly skilled at working with metal. It was a perfect job for Horace Smith, who had spent part of his youth helping to fashion weaponry at the biggest metal shop in the country: Springfield Armory.

Smith was born in late 1808 in rural Cheshire, Massachusetts, a town best known for having presented President Thomas Jefferson with a 1,235-pound piece of cheese produced by local dairy farmers, who were pleased that their favored candidate had beaten John Adams in the 1800 presidential election. (Cheshire was the only town in Berkshire County to support Jefferson.) Newspapers from around the country gleefully tracked what came to be called the Mammoth Cheese as it journeyed from Cheshire by wagon and boat until arriving in Washington for presentation to Jefferson on New Year's Day 1802. It was served at the White House and elsewhere for several years until the inedible remnants were disposed of, possibly in the Potomac River.

Although early nineteenth-century Cheshire was starting to show signs of industry with mills and tanneries, Horace Smith's carpenter-farmer father, Silas, decided to seek better opportunities elsewhere. Around 1812 Silas and his wife, Phebe, gathered up their four children—Horace and his siblings: older brother William and two younger sisters, one an infant—and traveled forty-odd miles southeast to Springfield. There Silas took a job at the Armory in what were called the Watershops, where the heavy work was done as iron wheels harnessed a tributary of the Connecticut River to grind and polish metal into gun parts.

The Smith family arrived in Springfield at the right time. The War of 1812 had begun, with the country needing new weapons quickly and as many worthless old ones as possible resurrected by skilled hands. The Armory superintendent who took over in 1805 started imposing division of labor, a move that increased both productivity and tedium. By 1812 musket production had nearly tripled to ten thousand per year. With increasing ferocity, America's Industrial Revolution moved rapidly ahead at Springfield Armory. New state-of-the-art milling machines were installed that allowed workers to fashion parts faster with more precision but with numbing repetition. An inventor named Thomas Blanchard at the Armory created a lathe that would duplicate a gunstock, a revolutionary machine that sped up production of guns and was later a model for making standard shoe lasts. In the years right after the War of 1812, the Armory also pioneered the use of trip-hammers for welding gun barrels. Too many guns had performed poorly during the war, so Armory supervision was put under the Ordnance Department's tight control to increase efficiency.

With all this progress came noise at a deafening pitch that would continue to rise in the coming decades. One observer called it "a sort of rattling thunder" that kept men working at the hammers, unable to do anything but the job at hand, "for such is the incessant and intolerable clangor and din produced by the eighteen tilt hammers, which are continually breaking out in all parts of the room, into their sudden paroxysms of activity, that everything like conversation in the apartment is almost utterly excluded." The clamor, combined with showers of sparks scattering in every direction, "produce[s] a scene which quite appalls many a lady visitor when she first enters upon it, and makes her shrink back at the door, as if she were coming into some imminent danger."

Silas Smith was paid adequately, but money was still tight for him, as well as for the country, which was on the brink of bankruptcy when he began work at the Armory. Making things tighter for Silas was the arrival of a new Smith in 1815: a second brother for nine-year-old Horace and another mouth to feed. Soon William, the oldest Smith child but still a teenager, joined his father at the Armory to help the family bring in more money.

After he turned sixteen, Horace, too, became part of the Armory workforce. The Smiths were now a Springfield Armory family.

First Horace assisted in forging bayonets, lots of them, typically more than five hundred every month. This required more than sharpening the point on a piece of metal. Once a bayonet was finished, every part of it was measured to make sure it had the precise form and dimensions the military required. Bayonets had to fit in scabbards, so each one was measured with a scabbard gauge. The next step was hanging a weight on the bayonet's point to test its firmness. Finally an inspector would place the point on the Armory floor, where he bent it to see how elastic it was. If the bayonet had been tempered too high, it would break. Too low, it bent and stayed that way. In either event it would be condemned, and the man blamed for making a substandard bayonet would pay for the loss.

Holding workers personally responsible for poor products became a rule throughout the Armory. "It is immaterial whether the misfortune in such cases is occasioned by accident, or carelessness, or want of skill. In either case the workman is responsible," according to one account of operations at the Armory. What's more, "He loses not only his own pay for the work which he performed upon the piece in question, but for the whole value of the piece at the time that the defect is discovered. That is, he has not only to lose his own labor, but he must also pay for all the other labor expended upon the piece, which through the fault of his work becomes useless." Knowing this, workers had to pay close attention to what they were doing despite the tedium. If they didn't, they would pay—literally.

Through hands-on experience, young Horace learned how to make metal behave. After four years and thousands of bayonets, he started moving up. A dogged man rather than a bold one and shorter than average, the compactly built Smith walked erect, possessing "a patient energy which has had few equals," according to a newspaper account. It was said of him that his "mouth shows decision of character, and the whole expression of the face is that of a man always perfectly clear in his plans, decided in his purpose, and entirely amiable and kindly in all his associations in life."

In the spring of 1829, Horace Smith married local girl Elizabeth ("Eliza") Foster, who was three years his senior. By the end of the year the couple had a son they named Silas after Horace's father. Horace had moved up at the Armory to straightening and boring barrels, hundreds of them each month. From there he was assigned to mill various other gun parts while on occasion told to return to bayonet forging.

Things were going ploddingly well for the new family until 1831 neared its end in an unusually cold December for Springfield. It was a busy time for Smith, who spent half that month milling gunlock plates and the rest of it milling parts of flintlock mechanisms: 500 sears and 1,227 tumblers, internal pieces that were essential to make the guns shoot. Child mortality back then was common, and on Christmas Eve, it hit Horace and Eliza Smith hard. Their two-year-old son, Silas, died.

Grief did not interrupt Smith's work at the Armory. He spent twenty-six of January's thirty-one days milling lock plates. For the next ten years he remained an Armory employee, eventually earning the coveted title of "machinist," making tools and inventing several types of gunmaking machinery, including one that cut checkered patterns in rifles' hammers to make them easier to grip. During this time his elder brother William was promoted to supervisor, their father continued working on the Armory floor, and Eliza gave birth to another son, Dexter, before she died in 1836. Two years later Smith remarried.

Through the years Horace Smith worked there the culture of the Armory was changing. Early nineteenth-century workers, on the farm or in craft enterprises, generally governed their own time on the job. They were paid by the task rather than the time spent performing it. They could drink on the job and socialize, even in most early industrial factories, and the Springfield Armory was no exception. But things had tightened up before Smith left the Armory in 1842, especially the previous year when a tough, stiff-necked Army major named James Wolfe Ripley took over as superintendent. "Vigorous, assertive, stubborn" Maj. Ripley "expressed in his personality the quintessence of military precision and discipline." So dedicated was Ripley to procedure and regulations that as a young procurement officer he had refused a requisition for supplies during the Creek War because it had not been properly submitted. Word of this reached Andrew Jackson, the commanding general, who immediately threatened to haul Ripley into camp and hang him from a tree if he didn't release the supplies. Ripley relented. He brought to Springfield Armory the same iron hand that only Andrew Jackson could bend. Ripley quickly began making the place work more efficiently while ruffling feathers. In a few years, when Ripley was in a higher post, some gun barons would come to hate him.

Before Horace Smith settled in Norwich to work for Allen & Thurber, he had been employed by several firms, including fifteen months at Eli Whitney's gun company. The difference in Norwich from all the other places Smith worked was that for the first time he had a hand in making a multi-shot firearm.

At close range a pepperbox was imposing. Facing its clump of barrels would cause even the toughest villain to shake with fear. A major drawback, though, was its inability to be accurate. In *Roughing It* Mark Twain described George Bemis, a supposed traveling companion who had armed himself with a pepperbox:

> He wore in his belt an old original "Allen" revolver, such as irreverent people called a "pepper-box." Simply drawing the trigger back, cocked and fired the pistol. As the trigger came back, the hammer would begin to rise and the barrel to turn over, and presently down would drop the hammer, and away would speed the ball. To aim along the turning barrel and hit the thing aimed at was a feat which was probably never done with an "Allen" in the world. But George's was a reliable weapon, nevertheless, because, as one of the stage-drivers afterward said, "If she didn't get what she went after, she would fetch something else." And so she did. She went after a deuce of spades nailed against a tree, once, and fetched a mule standing about thirty yards to the left of it. Bemis did not want the

> mule; but the owner came out with a double-barreled shotgun and persuaded him to buy it, anyhow. It was a cheerful weapon—the "Allen." Sometimes all its six barrels would go off at once, and then there was no safe place in all the region round about, but behind it.

Travelers back then tended to go armed, many with an Allen tucked in a belt or under a pillow at night. Some of these pistols were personal weapons carried by soldiers in the Seminole Wars, but they were useful only at close quarters, when fast shooting was crucial and aiming was not. That's why riverboat gamblers had them; quick response at the card table was required when being called out for cheating. That pepperboxes lacked the accuracy of the fine single-shot pistols duelists used to defend their honor supposedly didn't deter two women in Buffalo, New York, from deciding to settle their differences with Allens, despite a state statute outlawing duels. The damage those two pepperbox pistols would have caused to all assembled can only be surmised, because both would-be combatants were arrested before they could open fire—or so goes the story.

At Allen & Thurber, Horace Smith was in the company of creative men. Ethan Allen had the pepperbox patent, while Charles Thurber was a mere brother-in-law with money and connections. Thurber might seem an unlikely gunmaker. He was a cultured man who had taught school before teaming up with Allen in the gun business. In his spare time, Thurber wrote poetry and composed hymns, two of which were sung at the July 12, 1844, consecration of Norwich's Yantic Cemetery. But he also had a sense of the mechanical. In 1843 he patented an early typewriter for use

> where writing with a pen is inconvenient by reason of incompetency in the performer. It is specially intended for the use of the blind, who, by touching the keys on which raised letters are made and which they can discriminate by the sense of touch, will be enabled to commit their thoughts to paper. It is intended for the nervous, likewise, who cannot execute with a pen. It is useful for making

public records, as they can be made with this machine as accurately as with a common printing-press.

Thurber's machine was the first with paper on a roller, which gave it "longitudinal motion with provision for accurate letter and word spacing." Although groundbreaking, the slow, crude device was never manufactured.

Horace Smith would leave Allen & Thurber by 1848 to start his own gunmaking firm. Fifteen years later Smith would again have dealings with Ethan Allen. This time, however, the two men would be on opposite sides, battling in court over a gun patent, the onetime teenage bayonet forger's assistant having become the lead partner in a company called Smith & Wesson.

6

THE INDENTURED BROTHER

If he hadn't gone into guns, Daniel Baird Wesson might have been a good shoemaker. His father, Rufus, thought so anyway. He had urged the boy to take up that line of work, which two of Daniel's four brothers did. Shoemaking had become a thriving business in Worcester, Massachusetts, the Wessons' hometown, ever since skilled European craftsmen arrived there before the Revolution. After a canal opened in 1828 and the Worcester and Boston Railroad opened seven years later, the town turned into a transportation hub with vibrant manufacturing. By 1837 the surrounding county, where good soil made farming the main industry, had thirty towns and cities manufacturing boots and shoes, making the footwear business a natural occupation for anyone, especially the Wesson men, all of whom made things for a living.

Following his brothers' example, teenaged Daniel tried working in a shoe manufacturing firm but soon found the job didn't suit him, so he returned to school and helped his father, who was not a shoemaker but was well known in the region for making superb wooden plows whose carved convex curves "found high favor" in the region. When cast iron began replacing wood in plows, Rufus gave that up to become a farmer, acquiring large real estate holdings. For a while, Rufus was a highway surveyor and tax collector.

Daniel's preference for firearms over laces and soles or farming made him easily influenced by another brother whose fondness for guns was akin to his. Edwin Wesson, thirteen years older than Daniel, was making a name for himself as a craftsman responsible for some of the finest target rifles one could buy. (Edwin insisted that they were *the* finest.) His shop in Northborough, only about ten miles from Worcester, was doing good business. Maybe the boy could be useful there.

It had not been an easy start for Edwin after his own apprenticeship to a gunmaker. When he launched his business in 1839 in part of what used to be a carding factory, he was strapped for funds. "The truth is," he wrote in a blank ledger book, "I got so little cash tis not worthwhile to keep a book on purpose." Then with a touch he found humorous, Edwin went on, "Therefore I transmografy [*sic*] you into an Order Book with your permission."

There were gunsmiths around producing rifles that could adequately do what was required of a long arm, but the ambitious Edwin Wesson wanted his to do more. In an era when target shooting was popular, Wesson decided to make the surest, most precise rifles available, ones that could cluster bullet holes close together from a distance, closer than any that came from guns made with less skill and care. Rifling—spiral grooves cut inside the barrel—was known to make a gun more accurate than one whose bore was smooth. Edwin thought that increasing the speed of rifling's twist from the breech to the muzzle—called "gaining" twist—would further enhance accuracy, so that's what he did.

To make his guns even more accurate and bring in extra money, Edwin teamed up with farm boy turned professional engraver turned professional miniature-portrait painter Alvan Clark from Boston. Clark was a competitive shooting enthusiast (then called "a prize shooter") who considered himself "the best rifle shot in the world." He did suffer from a jittery disposition, an odd trait for a marksman. Clark was "so nervous that he could scarcely bring a cup of water to his lips without spilling it. His hands would tremble until he began to take aim, and then they would be as stiff as a vice until the charge was fired."

Like Wesson, Clark possessed a meticulously inventive mind and had come up with an idea to make the finest rifles even more accurate. His creation was a false muzzle to put on the end of the barrel. This would protect the real muzzle from being malformed when the shooter shoved in the bullet and the cloth patch wrapping it. So good was Clark's brainchild that he got it patented and gave Edwin the exclusive right to make it part of Wesson rifles for the price of two dollars apiece.

Less than a year after his false muzzle was patented, Clark published "On Rifle Shooting" in *The American Repertory of Arts, Sciences, and Manufactures,* touting his achievements in devising the best combination of sights, projectiles, loading equipment, and, of course, his false muzzle. He also singled out Edwin Wesson for praise: "In point of performance, this gentleman's rifles have a reputation of the highest order, and an uncommon elegance in external form and finish." If anyone doubted him, Clark advertised, he was prepared to compete against the most skilled marksman to prove that a Wesson gun was second to none.

With that Edwin's reputation went national.

Alvan Clark and Edwin Wesson quickly became friends, sending each other letters about riflery, constantly considering ways to increase accuracy. "I must come soon and spend a day or two with you in experiments," Clark wrote his partner in rifle making in the summer of 1842. Occasionally in a letter to "Friend Wesson" he drew circles on paper in exact proportion to bullet holes in targets he had shot. In one the widest spread was three inches, impressive for the distance between muzzle and target but not good enough for Clark unless he could be consistent: "5 shots at 100 yards with the gun I had at your place but it will not do so all the time. It bothers me yet." Clark was even tempted to make rifles himself full-time—he had already tried his hand at the craft—but told Edwin, "I have about $100 worth of [portrait] work before me and I cannot make my friends think it best for me to turn rifle maker at present. I think I must put off any further thoughts of it until next spring at least." He would stick with miniature portraits and shooting.

Clark kept his promise to square off against anyone tempted to oppose him in a shooting match. When a man named Lumas questioned in print the muzzle inventor's claim of superiority, Clark wrote Edwin Wesson that he had challenged the man to "shoot against me at rest ten string shots at 100 yards for $500 using an old fashioned rifle against my improvement. This I know I can win. I should be pleased to receive a visit from Mr. Lumas." Lumas never picked up the gauntlet.

Other gunsmiths liked Clark's false muzzle. Too much, in fact, because they began making them in violation of Clark's patent. "Extensive infringements have been committed, for which I am suffering," he wrote in a sporting magazine in April 1842. Edwin Wesson was the only authorized manufacturer, Clark announced, threatening to prosecute every infringer "to the utmost extent of the law." With orders coming in for rifles, thanks partly to Alvan Clark, and the need to join forces with the muzzle inventor in his fight against patent infringers, it was the right moment for Edwin to bring Daniel into the business.

"Daniel likes to hunt but he had rather be at work in the shop on gunlocks or springs or something of that kind," Rufus Jr., one of the two shoemaking Wessons, wrote to his brother Edwin in the fall of 1842. The family agreed: Edwin would bring seventeen-year-old Daniel on board as an apprentice starting that December.

This brother-hiring-brother agreement was not a casual one. Like many master-apprentice relationships at the time, Daniel allowed himself to be formally indentured to Edwin until he turned twenty-one. The contract signed with legal preciseness on December 12, 1842, by both men as well as patriarch Rufus Sr. and formally witnessed by their sister, Jane, spelled out everyone's duties. Edwin was to pay Rufus $250 in installments spread out over several years. He also agreed "to house and clothe said Daniel B. Wesson in a suitable manner" and give him three months of "regular schooling" every year. In turn Daniel agreed to work hard for his big brother as he learned the trade. If Daniel got sick, their father would cover the costs and forgive part of the payments Edwin owed.

It was the perfect arrangement for both men. Their enthusiasm for fine guns was in complete harmony, as Daniel eagerly took to the business of making target and hunting rifles for discriminating customers in an era when craftsmanship was appreciated. By 1844, after two years of doing and learning, Daniel was mastering the gunsmith's art.

And art it was, for Edwin Wesson rifles were known for elegance as well as accuracy. Edwin chose fine wood for his gun stocks, usually figured walnut, adding touches such as German silver ornamentation in floral scrolls or figures of animals, but nothing that would detract from the gentle sweep of the stocks on their downward slope from trigger guard to butt plate. Separate parts of the rifles, including places where wood met metal, fit together with exquisite exactitude. Gracefulness was a bonus; accuracy was the selling point, inspiring buyers from well beyond Northborough to implore Edwin Wesson to make their guns.

Despite a shared love of guns and shooting and a reverence for being precise, the brothers had different temperaments. Though young, Daniel was measured, a modest man devoted to family and temperance. Edwin could be combative, yet the two men complemented each other. When Edwin had to leave Northborough, as the Clark patent battles frequently required him to do, Daniel minded the shop, gradually taking over aspects of the business. Edwin learned that he could trust Daniel with more responsibility than many apprentices received.

Daniel had passion for more than guns. He was about to turn twenty when he presented Cynthia Maria Hawes with a precious book—precious not because it had market value but because the giving was a courtship gesture. At the center of its elaborate leather cover, the golden figure of a young woman strode barefoot, spreading golden roses from a basket she carried in her left hand. Inside, all the pages were blank except for the first one, where, in elegantly precise but strong script, Daniel had written lines by Lord Byron beneath an inscription to Cynthia:

As o'er the cold sepulchral stone
Some name arrests the passer-by;
Thus, when thou view'st this page alone,
May mine attract thy pensive eye!

And when by thee that name is read,
Perchance in some succeeding year,
Reflect on me as on the dead,
And think my heart is treasured here.

One word differed from the original. For the last line, he thought "treasured" suited his sentiment better than "buried," which Byron had chosen.

It was February 28, 1845, and Daniel was in love. He would remain in love, as would Cynthia, but marriage was out of the young couple's reach. Her father, Luther Hawes, wanted a better prospect for his middle daughter. Hawes was a respectable house carpenter who owned a small farm yielding rye and Indian corn. He decreed that Cynthia deserved someone with more promise than a mere gunsmith who was not likely to support a family in the style he thought was his daughter's due. On this Luther Hawes was firm: The couple could court, but marrying Daniel Wesson was out of the question.

By the fall of 1845, Edwin Wesson had given his younger brother additional responsibility in the business. Besides working on rifles, Daniel spent his time making a pair of pistols just for himself—long-barreled, single-shot weapons with gently curving grips—and a case to house them. Edwin, while away defending Alvan Clark's muzzle from patent infringers, let Daniel run the operation. Edwin fretted about the business when he was absent. Daniel reassured him. "I think we are getting along with the work in the shop as well as could be expected," he wrote to his brother on November 30. Customers were coming in, Daniel reported, and he was tending to them. "Edwin," he soothed, "do not have so much anxiety about your things here as to prevent your enjoying your excursion, for I will take care of them the best I know. I was at home today. The folks are all well."

Although Edwin worried about his business—perhaps it was his nature to worry—he seemed to be prospering. Lead balls shot from his rifles were known to cluster tightly on distant targets. He could thank Alvan Clark in part for adding to his fame, but other shooters also contributed. A pair of riflemen publicly challenged "any two men in the United States to shoot off hand [standing], from 150 to 300 yards" on a $250 bet. The challengers warned that they would "use rifles made by Edwin Wesson" against any guns their opponents chose. Edwin's rifles were pricey, the equivalent of thousands of dollars each in today's currency, but shooters with money and a passion for precision happily lightened their wallets to own guns bearing the Wesson name.

Edwin's family also saw him as a success and relied on his generosity. In December 1845 his twenty-eight-year-old brother, Martin, asked for a loan of "fifty one hundred or one hundred & fifty dollars" to settle debts. Three years earlier Rufus Jr. had written him that "a little money would do me a great deal of good as I have got to use rather more than I have got or can get up here, and if it would be possible for you to help me a little, I should like it very much."

Edwin also needed money. Less than two months after brother Martin asked for a loan, Edwin signed a four-hundred-dollar mortgage on much of his gun-making equipment, including lathes, an anvil, a pair of bellows, and a variety of hand tools. With those as security, he promised to repay the loan on demand with interest. If Edwin had more orders, preferably from the military, and not just for fine, labor-intensive rifles commissioned by finicky customers, he could improve his prospects.

California attracted increasing American interest in the mid-1840s, including that of a company that was fitting out an exploring expedition and wanted to know whether Edwin could supply rifles plus a hundred large revolving pistols. Wesson had no pistols to sell, so he reached out to Allen & Thurber to see if he could be a middleman for that company's pepperboxes, allowing him "to make a small commission myself."

Within a year an opportunity to make more than a mere commission would come tantalizingly close.

7

REVOLUTION IN THE BLOOD

Month after month passes, year after year;
Thinking, Thinking, Thinking;
Jeered at and ridiculed, thought of as "queer";
Thinking, Thinking, Thinking.

And then the idea you've slaved for is struck,
Thinking, Thinking, Thinking;
Envy and indolence call it pure luck;
Thinking, Thinking, Thinking.

CHRISTOPHER MINER SPENCER, "THE INVENTOR"

At harvest time Josiah Hollister had a simple chore for his grandson: Take a load of potatoes to town and sell them. Then come home.

Although not yet a teenager in the early 1840s, Christopher Miner Spencer was up to the task. And he would do his best, at least in part because he respected his eighty-eight-year-old grandfather, who had told him at length of his service in the American Revolution. In fact Hollister had told everyone about it, and with understandable pride, not just because he was a veteran of what all saw as a glorious struggle for independence, but because he had personally known George Washington as his commander. With his own hands Hollister claimed to have built a tall wooden wardrobe for Martha Washington's clothes (he called her "Lady Washington"). People said she delighted in it. He always delighted in talking about it.

The old man's affinity for crafting and refining physical objects had long been recognized. While one account credits Hollister with good marksmanship, his main contribution to the fight for independence was something entirely different. Twenty years old when the Declaration of Independence was signed, the straight-spined, active, strong, and ambitious Josiah Hollister had joined the cause as an "artificer," someone who could create and then fix anything that required fixing, from wagons, to equipment the soldiers needed, to battlements, even to making tents and clothing. Under the immediate command of French and Indian War veteran Jeduthan Baldwin, who had shown engineering talent during the siege of Boston, Private Hollister summoned his craftsman's gifts in aid of the army dedicated to breaking the British hold on his homeland. With his fellow artificers Hollister served in several places, including Valley Forge, where he had time to create Martha Washington's wardrobe. When the war ended, Hollister came home to Connecticut, settled down on a farm, got married, and began a family. He would eventually father thirteen children by two wives.

Spencer, whose childhood nickname was "Crit," had come to live on his grandfather's farm at the age of eleven. His parents and siblings lived in town a short distance away, so exactly why he was to spend a few years with his grandfather is not completely clear, though it was probably to help the old man, whose second wife, Crit's maternal grandmother, died at the age of eighty in early June 1844. There is no evidence of serious family strife, and Crit's father, Ogden, a successful man in the wool business, was far from poor. It appears that the family may also have wanted Hollister to teach Crit something about farming. Although the man was old, he had sharp eyes and ears, and his memory was clear. He could look after Crit.

The Hollister farm was just outside Manchester, a town several miles due east of Hartford and not far from where the old man was born in an era far different from the boom years that followed. Originally Podunk Indian territory, the area was settled by Puritans who made it a farming community during the century that led up to the Revolution. By the mid-1840s Manchester was home to a half-dozen paper mills, two gunpowder mills, five woolen mills, two cotton mills, and two silk mills, as well as some other manufacturing enterprises. Industry had come in earnest. The mills, powered by water flowing along several brooks and the Hockanum River, which itself flowed into the long, languid Connecticut River, were expanding, while new ones were being built. Their wheels turned machinery that made an increasing variety of products, from glass to lumber to the town's traditional textiles. The growth of business in Manchester meant that neighboring farms had a healthy market for their produce. Hence Crit Spencer's potato-selling trip to town.

Once Hollister's potatoes were sold and the money safely pocketed, Crit started on the journey back to his grandfather's farm. Then the boy saw something novel moving through town, a colorful procession on its way to a hastily erected complex unlike the staid Manchester's mills or other supremely functional buildings. The circus had arrived, meant for pure fun and soon to vanish. The parade heading toward the tents was exciting, awe-inspiring, alluring to everyone, especially children and perhaps especially Crit, a boy with an abiding curiosity about new things, functional and otherwise. He convinced himself that his grandfather wouldn't mind if he watched for a bit.

A growing crowd followed the parade, sweeping up Crit Spencer as it went until it stopped at the tents. There, hawkers' cries and other sounds and sights unlike anything else in the industrializing town enticed many to become part of the experience. His grandfather didn't need him immediately, Crit reasoned. There was plenty of time before he had to return to the farm, so why not take advantage of the moment and explore the wondrous offerings now before him? The sun had set by the time the boy started for home. Crit had spent more than hours at the circus; he had also spent potato money—not all of it, but enough to convince Hollister not to entrust his grandson with another errand like that anytime soon.

Crit's costly circus detour would be but a small bump in the relationship he had with Josiah Hollister. Grandson and grandfather—nearly eight decades apart in age—were bound together by more than blood and build. (Although Crit would never be as tall as his grandfather—he reached five foot five by adulthood—they both retained lean but hardy physiques throughout their lives.) They shared an interest in building things and in making other things work better. They had a knack for it, perhaps even a passion, for when his mind focused on such tasks, Crit could not be diverted, as he had been with the circus.

By the time Crit Spencer went to live with his grandfather, the ranks of Revolutionary War fighters had dwindled to an honored few. During the brief period he lived under Hollister's roof—only a handful of years—Spencer devoted himself to the art of woodworking. His grandfather's reluctance to send him on more potato-selling missions did not keep him from trusting the boy to use an old foot lathe at the farm. And use it he did, regularly and often, so much so that his Aunt Harriet thought the boy's time should be spent on other, more productive activities. Better to chop wood, she would say, or do something practical rather than turn out fancy but useless things in wood. Hollister disagreed. He saw something worthwhile in Crit's labors, or at least a potential.

"Let the boy alone," he told Harriet, "and he may make something yet."

It could have been his grandfather's tutelage that helped inspire Crit, or perhaps inventiveness and industry were in his genes, and not just from the Hollister line. The Spencers claimed as an ancestor a respected furniture maker who had crossed the Atlantic as part of the Great Migration in the 1600s. Chairs built by Thomas Spencer were known for their sturdiness and pleasing lines. For decades his descendants would carry on the tradition.

Crit was interested in more than wood. Firearms lured him, too, and when Hollister eventually turned over to him the flintlock musket the old man had carried in the Revolution, the boy decided to improve it. Like most shoulder arms of the day, the gun's barrel was long—too long, as far as Crit was concerned—so he decided to fix it. Lacking the right tool for the task, the boy made his own. After taking a large knife from its sheath, he hacked grooves into its blade with an axe to create a crude saw. The now-serrated blade's metal was strong enough—and so was the boy—to lop off a section of the barrel. In Crit's mind, this made the weapon appealingly modern. His grandfather's reaction to the surgery performed on his musket is not recorded.

Josiah Hollister lived another five years beyond the summer of 1844. His longevity provided a crucial connection between the war that had given birth to the United States and the war that would tear it apart. Though he would not live to see it, his mentorship of Crit would aid the country in a conflict as monumental as the one in which he had served.

8

EYES WEST

War was his element, the bivouac his delight,
and the battle-field his play-ground,
his perfection and inspiration

J. JACOB OSWANDEL ON TEXAS RANGER CAPT. SAMUEL H. WALKER

Samuel Colt was in limbo—or, rather, as much in limbo as this inventor in constant motion would ever get. It was the end of 1844, and his revolver business was dead. His underwater mine, demonstrated so spectacularly, was going nowhere. Colt was thinking about the electric cable that he used to set off the mines. If his friend Samuel F. B. Morse could speed words through wires, as he had during the Democratic presidential convention in Baltimore that May, why couldn't he get in on the act? Fast-arriving business news could be a gold mine. Colt was convinced. This time he would finally become rich and famous.

In the meantime Colt's main hustle was waterproof tinfoil cartridges for guns, but this venture was not the cash bonanza he needed to resume gun manufacturing. He still had the patent for his revolver, and he was unshaken in his belief that it was a worthy weapon—or could become one, with the right improvements, the right endorsement, and the right circumstances. Over the next year and a half, all three would align for Samuel Colt. The United States hungered to expand. War was brewing with Mexico. And a patron saint of that expansion, a man lauded in military circles and expert with handheld weaponry, had just escaped death at the hands of Comanche Indians, thanks to the firepower of a Colt revolver. That was just the beginning of a collaboration that would bend the trajectory of the country.

Samuel Hamilton Walker was born in Toaping Castle, which was not, in fact, a castle. It was a log house made of white oak, in Maryland, a dozen miles northeast of Washington, DC. Built by his grandfather and great uncles, it was named after the family's ancestral stronghold in Scotland. Walker was raised at Toaping on tales of valor and rebellion. His ancestors had fled across the Atlantic with a bounty on their heads after failed attempts to kick King George I off the Scottish throne. Walker's grandfather, father, and uncles again fought the monarchy during the American Revolution. Their stories of heroic conflict thrilled young Walker, who later wrote that "nothing so much interested me as to read of the chivalry and noble deeds of our forefathers in the wars with Great Britain." He revered his father, Nathan, "always a true friend to the cause of freedom and justice," which were "principles he always endeavored to instill into the minds of his children."

Nathan apprenticed his teenage son to a carpenter, but there was no glory in chopping and sawing wood. So at the age of nineteen Walker volunteered to head south to fight Indians who were unhappy with Americans forcing them to move from their homelands. First he was a private helping the Army suppress a Creek uprising in Alabama. Then he was in the swamps of Florida, taking part in the Second Seminole War, a drawn-out, costly contest between whites eager to remove Indians from the region and Indians whose "cultural and spiritual identity was fundamentally linked to their continued presence in Florida. . . ." After his year's enlistment was up, Walker stayed in the South as an Army scout. Some say he also spent time as a surveyor and with an older brother ran both a hotel and a construction company in Florida, as well as a steam-powered sawmill that was eventually destroyed by fire. Despite having been part of an army at war, there had been no combat for him, nothing to prove his mettle. Instead he had been frustrated by the relentless, regimented drudgery of Army life and the behavior of some of those in ranks above him (as a private, most everyone outranked Walker), men he thought unworthy. In 1840 Walker risked burning the bridge behind him by publishing a screed against the military that thundered "no man with purely patriotic feelings can content himself to remain long in the United States army, under the present abuses of power which is daily practised in it." As for officers, Walker said many of them were "tyrannical, pitiful, cowardly, disgusting, and contemptible scamps. . . ."

The United States was fast ripening into a place where an incautious man could swashbuckle his way into battle. Walker was like much of the American nation itself in the early 1840s: young and bold, hungering to reach what it saw as its due greatness, to conquer the limits of geography and take its rightful position in bringing progress as far as dedication would allow. Part ideological, part religious, part nationalist, part racist, part bursting with raw youthfulness, both Walker and a large share of the United States population saw themselves as soldiers in the vanguard of destiny.

The 1840s was a turbulent time in the history of a turbulent country. The Panic of 1837, which contributed to the demise of Samuel Colt's gun company, and the ensuing depression heightened the desperation of people who feared that good times in the young nation had come to an end. Economic woes drove many to novel brands of Christianity that broke from rigidly traditional denominations and promised that, under proper guidance, true believers would find the path to glory. The Second Great Awakening, a frenzy of Romantic religious movements, had been cascading through the country. Revival meetings made devotees thirsty for social reform and salvation and weakened the grip of staid, old-fashioned Protestantism. The movements promoted virtue, temperance, hard work, and charity. Mormonism drew converts as well as enemies. Its prophet, Joseph Smith, was murdered by a mob in late June 1844, becoming the first candidate for US president to be assassinated. Followers of Millennialism believed Christ's second coming was at hand. "Religious insanity," linked to evangelism, ballooned at this time, with admissions to mental asylums in Hartford, Boston, and Worcester reaching a peak in the 1840s.

Nativism was also on the rise, buoyed by fear of the growing number of Catholic immigrants, mostly from Ireland. Violence reached a peak in 1844, when two bursts of rioting erupted in Philadelphia, one in May and another in July. In the first a procession of nativists marched through the city streets holding aloft a torn American flag that bore the painted words "This is the flag that was trampled under foot by the Irish papists." The *Native American* newspaper roared, "The bloody hand of the Pope has stretched itself forth to our destruction. We now call on our fellow-citizens, who regard free institutions, whether they be native or adopted, to arm. Our liberties are now fought for;—let us not be slack in our preparations." The furor was just as intense two months later. By midsummer, scores of people in Philadelphia had been injured or killed.

Despite the tempest in the country, it got back to business and recovered from much of the economic devastation wrought by the Panic of 1837. New railroad lines reached Springfield, Massachusetts, allowing the Armory to ship guns to more places more quickly. In 1844 a stationary steam engine was installed in the Armory. When it worked, which wasn't always, it relieved the factory from depending on the moods of rivers to run its heavy machinery. The Armory's buildings and grounds were being improved, thanks to Superintendent Ripley, who ramped up production while reducing costs at the same time.

Both American industry and the American soul were ascendant. A byproduct of these twin forces was a sense of "manifest destiny," a term first coined in 1845 to articulate a sense that the hand of Providence was directing the United States to spread its dominion over the continent, all the way to the Pacific Ocean. Many saw the country's westward expansion as part of God's plan for the young nation to establish a chosen civilization in "heathen" lands, regardless of the claim or rights of its first inhabitants. What had been a trickle of fur trappers and explorers along the Oregon Trail had grown into a stream, as white Americans headed west to create new and better lives in a promising climate. By 1845 five thousand people had made the journey. Many more would follow.

And then there was Texas. Expansionists, especially those in slave states, wanted the Lone Star Republic to become part of the United States. Mexico considered Texas part of its sovereign territory and vowed to oppose its being gobbled up by the Anglo nation to the north, by force if necessary.

Samuel Walker, at age twenty-five, saw Texas as a bold land out there, rich in glory, born in liberty, an oasis marking civilization's next step on its journey. The young republic, like the United States, was still fresh from wresting itself from what it saw as an old order of stultified hierarchy trying to rule from afar. Mexico had never accepted Texas's independence, so raiders kept coming from below a line that the republic considered its southern border. Indians, too, were sacking settlements and needed to be beaten back, as the settlers saw it.

"The love of chivalric immortal fame still clings to my heart" and "urges me on," Walker wrote to his sister-in-law from Florida in January 1842, shortly before leaving for Texas. He found his tribe in the Texas Rangers, who battled both Comanche and Mexican raiders, and who shared Walker's sense of risk and rebellion. No man among the Rangers wanted to kill Mexicans more than Sam Walker did. In late 1842 he joined a militia expedition against Mexican border settlements, partly in retaliation for a massacre of Texans by the Mexican army earlier in the year. When it became clear that the Republic of Texas would not back the expedition further, the officer in charge ordered his men to re-cross the Rio Grande and come home. But Rangers hated to shy away from a challenge, and just over three hundred of them, including Walker, disobeyed orders and went deeper into Mexico. They were eventually forced to surrender after encountering an immensely superior Mexican force.

The captured Texans were marched toward Mexico City. Fearing that they would never see home again, 181 men escaped in February 1843, but the harsh land forced 176 of them to surrender or be recaptured within a couple of weeks. Enraged that the Texans had escaped, General Antonio López de Santa Anna ordered that all the escapees be executed. Balking at such wholesale slaughter, another Mexican general had the order modified: only one in ten would die. To select which men would be shot, each Texan was blindfolded and told to pick one bean from a pot containing 159 white beans and 17 black. Black meant death. Although Walker pulled out a white bean, some of his compatriots did not. At dusk on March 25, 1843, the Mexicans shot seventeen Texans. The rest remained confined in various places, including Perote Castle, a grim, high-walled fortress of volcanic rock west of Vera Cruz built by the Spanish in colonial times to keep treasure safe before it was shipped to the mother country. Every so often some, including Walker, managed to escape.

"I write to inform you I am in good health, hoping this may find you the same," Walker wrote from prison to his sister-in-law, six weeks after what came to be known as the Black Bean Episode. "When I shall return I don't know as my experience thus far has only increased my anxiety & ambition to fight the mexicans. I have witnessed the murder of 18 of my comrades in cold blood and I am determined to revenge their death if I have an opportunity."

Two months later Walker escaped, making his way back to Texas, where he joined the Rangers. He would get his chance at revenge.

Early in 1844 the administration of John Tyler, then US president, joined the administration of Texas president Sam Houston to hammer out a treaty that would bring the Lone Star Republic into the union. But the Senate overwhelmingly rejected the treaty at the very moment that the Texas Rangers—including Jack Hays and Samuel Walker—defeated Yellow Wolf's Comanche band in the Hill Country. The fate of Texas, however, remained a live issue, becoming a focal point that autumn in the presidential election.

Whigs and Democrats had selected their nominees in the months leading up to the election, and each had a massive, energized following. The Whigs' choice, as expected, was Kentucky's Henry Clay, a former US senator and secretary of state who had long been considered, particularly by himself, to be perfectly suited for the presidency. For their standard-bearer, the Democrats picked James Knox Polk, a slave-owning former speaker of the House and Tennessee governor. Clay and Polk had vastly different visions for the country. Whigs wanted the United States to develop industrially and economically within its borders, with the help of the federal government. "It is much more important that we unite, harmonize, and improve what we have," Clay wrote to a fellow Kentucky politician, "than attempt to acquire more."

Democrats wanted to expand the United States westward by kicking the British out of Oregon and annexing the Republic of Texas. They had a true champion in Polk, a protégé of Andrew Jackson and an industrious and forceful politician with a humorless disposition. He would push hard to acquire Texas, regardless of the Senate's recent stonewalling.

Votes were counted in December, and Polk eked out a victory. Clay might have won if a third-party, antislavery candidate had not siphoned votes from the Whigs. The mood in Congress was changing, and the winds were shifting toward annexation of Texas. War with Mexico was approaching inevitability.

On February 28, 1845, four days before his inauguration, President-elect Polk got his wish: Congress agreed to annex Texas. Nearly eight months later, Texas voters approved a constitution paving the way for their republic to become a US state. By the spring of 1846 opposing armies had gathered near the Texas–Mexico border, though a universally recognized boundary did not exist. The United States claimed that the Rio Grande was the demarcation line. Mexico said it was the Nueces River, as far as 140 miles north of the Rio Grande. The difference would lead to the first battles, and a declaration of war by the US Congress on May 13, 1846. With that the United States began a bloody imperial sweep westward—and Samuel Colt began his own campaign.

Colt wanted to sell his guns to the Army. Two months after the war officially started, he wrote to President Polk asking for a commission to lead a rifle regiment "based upon the fact that I have spent the last ten years of my life with out proffit in perfecting military inventions. . . ." Colt had seen news reports that Walker had turned down an invitation to take command of a regiment, so he offered himself as a suitable replacement. If the president wished, Colt added, he was ready to make his case in person. But Polk did not give the inventor, who had no military experience, what he had asked for. The man behind the revolver needed to find another way to get his guns enlisted, so he turned to the Texas Rangers, who were joining in the war against Mexico and had used the first incarnation of his revolving pistol. Colt hoped they might provide a winning endorsement.

So eager to get into the fight was Samuel Walker that he had signed on as a private in General Zachary "Old Rough and Ready" Taylor's volunteer army months before war broke out. He was with Taylor in the battles of Palo Alto and Resaca de la Palma the previous May and scouted for the army that summer, often slipping behind enemy lines to bring back information on enemy strength. In one engagement Walker's horse was shot while he was in the saddle. When a Mexican lancer bore down on him, Walker pulled himself from under the dead animal, shot the lancer, grabbed the reins of the man's horse, and swung himself into its saddle for more fighting. (Admiring citizens of New Orleans bought Walker a replacement horse named Toronado.) At Monterrey in September, Walker and his fellow Texas Rangers used their skills at guerrilla tactics against Mexican forces in the city, teaching urban warfare to United States Army regulars. In his official reports Taylor singled out Walker by name as having "performed very meritorious service as a spy and partisan."

Despite occasional brushes with authority, which included being tied to a tree as a private in Florida for disrespecting an officer he considered an immoral drunk, Walker was all military. He was generally low-key during down times, but when the fight was on, so was he, becoming "rapid, untiring, terrible," a man who in those moments "seemed to change his whole character and appearance, and arise a new being, entirely superior to himself." It was said that there was no fear in Walker, for he "seemed to seek danger; to love it for itself."

Except for their blue eyes and same first name, Samuel Colt and Samuel Walker were very different men. Colt was tall and brash, charismatic and confident, always pushing himself forward. Walker, on the other hand, could be "quiet" away from the battlefield to the point of being uncommunicative. With a slender frame, slightly rounded shoulders, and a young, seemingly soft face, Walker did not look like the fighter he was. While Colt had a fondness for alcohol and cigars, Walker was "exemplary in his habits, and used neither liquor nor tobacco in any form."

Most important of all, Captain Walker had the military bona fides—and the warrior mystique—that Colt lacked. When he came back East that November to recruit men for his company of mounted soldiers, his reputation preceded him. Walker rejected the idea of a formal ball held in his honor but accepted an invitation to attend a gathering at an Odd Fellows hall in Washington. There, people stood in line to praise him, one by one, as a patriot and soldier. Three days later President Polk, who had appointed Walker captain in the mounted rifle regiment, received him for a visit—an honor denied Colt.

Earlier in the year Colt had written to Jack Hays to promote his revolver dreams but had not received an answer. Now Colt pivoted to his subordinate superstar.

"I have so often herd you spoaken off by gentlemen from Texas that feel sufficiently aquainted to trouble you with a few inquires regarding your expereance in the use of my repeating Fire Arms & your opinion as to their adoptation to the millitary service in the war against Mexico," Colt wrote in a letter he had hand-delivered to Walker after the captain's arrival in New York, the inventor's promotional bravado eclipsing his ability to spell. He had heard tales of Walker's and Hays's exploits, Colt explained, and had "long desired to know you personally. . . ."

The trouble with the modern military, Colt told Walker—perhaps preying on the man's aversion to Army hierarchy—was that ordnance officers shied away from new tools, even ones that worked well. Colt pleaded for Walker to describe his triumph in the context of weaponry and suggested that he urge the president and the secretary of war to equip their men with Colt revolvers. The design was improved now, Colt claimed, but the gun could still become "the most complete thing in the world"—if someone like Walker would further advise.

Walker took the bait. In a letter to Colt on November 30, four days after his audience with Polk, he wrote: "The pistols which you made for the Texas Navy have been in use by the Rangers for three years, and I can say with confidence that it is the only good improvement [in weapons] that I have seen. The Texans who have learned their value by practical experience, their confidence in them is unbounded, so much so that they are willing to engage four times their number." Walker told Colt of the battle with the Comanches, which Colt already knew about. "Up to this time these daring Indians had always supposed themselves superior to us, man to man, on horse. . . . I can safely say that you deserve a large share of the credit for our success. Without your Pistols we would not have had the confidence to have undertaken such daring adventures."

There could indeed be improvements, Walker advised. With the help of an old gunmaking friend, Colt had already been tinkering with the Paterson design to eliminate at least some concerns the military had about its reliability. Together Walker and Colt came up with something they thought would give any soldier an edge. It was a behemoth of a handgun that pushed bigger lead balls with greater force than did the old Patersons while using the same patented method of revolving the cylinder and locking it in place for each shot. Instead of five bullets, the new pistol's cylinder would contain six. Gone was the curious trigger that was hidden in the gun's frame until it dropped down when the shooter cocked the weapon. In its place was an exposed trigger protected by a metal guard attached to the frame. And this time a toggled lever below a nine-inch barrel allowed the revolver to be reloaded without having to take the gun apart; the standard Paterson had required dismantling, so that a full cylinder could be swapped for a spent one. Each mounted soldier could carry two of these revolvers in holsters draped across the front of his saddle, one on each side. Such firepower, twelve man-killing shots in a row, in the hands of a single rider was unheard of.

On December 7 Chief of Ordnance George Talcott gave Colt thrilling news: The secretary of war wanted one thousand revolving pistols, pronto. Colt had three months to deliver. That same day, Walker penned a letter to Colt:

> I doubt not you will receive orders for at least two thousand more in a very Short time, In case you should not however do not let it prevent you from making at least that number as I feel confident that you can sell that number in New Orleans & Texas at a rate that will pay well. . . . [Y]ou have nothing to fear in regard to their Success and allso of their being brought into general use by our Mounted light troops which are indispensable to the protection of our vast & extensive Frontier I think you would do well to turn your attention entirely to the manufacture of those arms feeling perfectly satisfied that you will be well rewarded.

The elated Colt was back in business. Unfortunately he had neither machinery nor a factory. A three-month production deadline was daunting, but for Colt there was no option; he had to agree. The next day he wrote to the arms maker Eli Whitney Jr., son of the cotton gin creator, in New Haven, Connecticut, and asked if the Whitney machinery could make his revolvers.

Beneath his signature, Colt added a request: Did Whitney remember the name of a gentleman who, two or three years before, had shown specimens of a steel rifle barrel in Washington to get a government contract? Colt wanted to know who and where he was. Whitney's response was immediate. He had lots of work to do for the government but could take on Colt's project if there really was an order for revolvers. And he knew about the man whose identity Colt was after.

"The name of the person you enquire for," Whitney wrote, "is Mr. Remmington [*sic*] of Remmington Herkimer Co. N.Y."

While Colt was negotiating with Whitney, he kept up a steady correspondence with Walker, partly to enlist the captain's aid in making sure that things went smoothly with the government, but also to keep him happy in case production didn't go as promised.

"I have every reson to believe that I can complete the order by the time spesifide & if it is within human power to do so *it shall be dun,*" Colt wrote Walker on December 10, hopeful that rigid military minds wouldn't fuss about design changes or slow down production with pointless demands. Then Colt turned to fawning praise. His revolvers owed their reputation to Walker and Hays, Colt reminded the Ranger, and he would never profit by placing in their hands a gun that would not do them credit.

Colt decided not to strike a deal with Eliphalet Remington for gun barrels to be shipped to Whitney's New Haven factory, where workers were to fashion the novel weapon's complex mechanism and put all the parts together. But he did reach out to another man who had a sharp talent for making fine gun barrels. On Christmas Eve 1846 Colt sent the man a letter.

"You may expect to see Captain Walker U.S. Rifles & myself in Northboro on Tuesday or Wednesday next," Colt wrote to Edwin Wesson.

9

CRAFTING A WARRIOR'S TOOL

Edwin Wesson was relieved when Captain Samuel H. Walker walked into his shop. Here was the kind of customer who could save the family operation from financial peril. Walker was no weekend target shooter looking for a single prestige rifle, like many of Edwin's customers, but a warrior fresh from the battlefields of Mexico. If Walker could persuade the military to place a bulk order for rifles from Edwin, the Wessons could bail themselves out of the four-hundred-dollar mortgage on their gunmaking equipment.

Samuel Colt was not with Walker when the captain arrived at the Wesson shop, but he made his wishes known in writing. He chose not to ask Edwin and Daniel to make barrels for his new revolver. Instead Colt told the Wessons that he wanted molds into which molten lead would be poured to harden into bullets for the new guns. He also asked for a set of tools to smooth out the surfaces of the lead balls once they came out of the molds.

In a February 5, 1847, letter he wrote:

> I want you to bear in mind that it is necessary for my balls to have a little taper or curve at the lower end so that they will enter the cylinder by the pressure of the ramrod or lever without shaving or cutting either one side or the other, & the balls when in should be sufficiently straight on the side to bind in the chambers so as not to start by the recoil.

With his three-month production deadline looming, Colt was in a hurry. "I want these things as soon as you can possibly make them," Colt told Edwin. "You can send one or the whole at a time as your earliest convenience suits.

As far as Walker was concerned, it didn't matter who made the barrels or molds, as long as Colt's revolvers came to his troops on time. What Walker did want from Edwin was what the target master excelled at: a personal Wesson rifle, weighing under eight pounds, made to Walker's specifications. The captain told Edwin that his brother, Jonathan, also wanted a rifle made for himself.

The brothers Walker and Wesson had more in common than an interest in rifles. They shared a respect for craftsmanship. Walker's father had passed down to his sons both the ethic of hard work and the skill to make things, "so that we are all mechanics, a distinction which I am proud to acknowledge," Walker once wrote, showing his distaste for snobbery and those wielding undeserved authority. "I am aware that aristocrats and lovers of power and tyranny will turn their noses up, as the old saying is, to my remarks, but I care not for that, as my wish is to warn the workingmen of the country to watch with unceasing vigilance every movement of those in power, who, instead of being public servants as they should be, would rather make the public their servants." Walker appreciated a man who worked well with his hands, especially if those hands made exceptional weaponry, and so he was keen on commissioning a personal item from Edwin Wesson.

Edwin was delighted to build rifles for the Walkers, but he wanted something for himself. If Colt wasn't going to give the Wessons a big barrel contract, perhaps the captain could convince the military to adopt their rifles. Edwin knew, as did Colt, that Walker's opinion carried great weight with the military. And the captain had all the right contacts, including the president of the United States. Besides, Walker knew how guns should work, and Wesson rifles worked splendidly; there could be no better endorsement than the captain having ordered one for himself. Certainly, Walker told Edwin, he would gladly promote the family's handiwork. And he moved quickly.

"I have got the matter before Congress and it is in the hands of the military committee," Walker wrote Edwin from Washington on January 18, 1847. "All seem favorable to such arms & equipments as I recommend. As soon as there is any action I will let you know all about it." Walker also showed Chief of Ordnance Talcott a target shot through with bullets from a Wesson rifle as evidence of their accuracy. By the end of the first week of February, Walker had left Washington for Baltimore, where he continued recruiting for the military. Being a native Marylander as well as a national hero led to more young men signing on with him. Edwin Wesson was hard at work making Walker his rifle, knowing that it would be a calling card in Washington's circles of influence.

"I think I may yet secure an order for enough at least to arm our Regiment," Walker wrote to Edwin. "I expect Colts Pistols here by Friday next and would like to receive the Rifle at the same time or sooner."

Texas senator Thomas Jefferson Rusk, who had helped avenge the Alamo at San Jacinto and now backed war with Mexico, lobbied the Senate to consider Edwin's rifles. "I am certain of enough to arm my company at least," Walker wrote to Edwin, "and I wish you would commence one hundred without any delay, as I have no doubt of this much."

Maryland politicians also did their native son's bidding. On February 24 the state legislature passed a resolution lavishing praise on Walker for his feats of "personal daring and bravery," noting that more than a hundred Marylanders, "implicitly relying upon his unsullied reputation for chivalry and courage, have voluntarily chosen to follow him to a triumphant victory or glorious death. . . ." Because these green soldiers needed to have trustworthy weapons when they went to war, the legislature asked its Washington delegation "to urge upon the General Government the necessity and propriety of furnishing said company with such rifles as may, in the opinion of Captain Samuel H. Walker, be most suitable for the service on which said company are about to be engaged." Edwin Wesson's future seemed brighter than ever.

In a letter written on the day of the Maryland legislature's resolution, Walker brought Edwin good news and bad. "I am much pleased with my

Rifle," he wrote. He had "not had time to shoot it as yet but am perfectly satisfied that it will suit me." Two other men in his company also wanted Wesson rifles. So did a sergeant, who asked Walker to enclose forty dollars for one exactly like his captain's.

Here was the bad news: "[I] used every exertion that was in my power to secure the [military] order which I contemplated and am sorry to inform you that I was unsuccessful." Walker offered no further explanation, at least not in that letter. For whatever reason, the chance for the Wesson firm to become a mass-producer for the Army was gone. Edwin still had hope for Walker's company of volunteers. If the gallant, celebrated captain had pull for anything, certainly it would be to supply the men under his personal command with good rifles.

But now Walker's mind was on Colt's revolvers. Soon he would return to the battlefields, where he wanted—where he *needed*—firepower from the multi-shot handguns he had helped design. On March 6 Walker arrived by train at Newport Barracks on the Kentucky side of the Ohio River across from Cincinnati. He had tired of recruiting in the cities, with all the hoopla and the coddling by parents of young would-be soldiers, and wanted to get ready for battle. There were fine men in his company, he told Colt, "and all I want now is the arms, I hope you will rush things as rapidly as possible and get me enough completed for my company before I am ordered to leave, write and let me know the earliest possible moment you can get them ready, there is nothing now not even a Female that gives me so many thoughs [*sic*]."

Two more weeks passed, and still no guns. *Where were they?*

"[D]o for heavens Sake rush things as rapidly as possible and send me some of the Pistolls immediately," Walker begged Colt in a hastily written letter from the barracks. "I want to commence drilling my men on horse back with them," he told his co–gun designer. "[E]very thing now is depending on you," Walker stressed, asking to hear from Colt "immediately *if not sooner* and let me know when they will be forthcoming, find out how things are likely to work in relation to inspection &c and let me know all about it." Maybe Eli Whitney could speed up production by putting all his workforce into making the revolvers, he suggested. If the factory managed to turn out enough guns to arm his detachment before they left for battle, Walker told Colt, "you shall wear the brightest Laurel of our first victory and the Glory shall all be thine."

Colt was having problems with the Ordnance Department. Colonel Talcott and his subordinates were prejudiced against his revolvers, Colt complained in a March 21 letter to Jack Hays, who was then in Washington. Perhaps Hays could push things along. Three days later Walker wrote Colt yet another letter, pleading for revolvers "as rapidly as you can posibly" make them. Even a small amount, say, only twenty at a time, "would be desirable, dam the odds about all inspection send them to me and I will inspect them, and make it all right."

Colt had every reason to think that his revolvers would not be made on time. In addition to frustrations from dealing with Ordnance, he had production problems. He certainly wasn't going to make things worse by telling Walker not to expect the guns soon, especially when rumblings in the Army led the captain to think he was about to be sent into action. "[I]t will never do for me to go into another engagement without some of these Pistols."

Walker was right. "We are now on our way to the seat of War," he wrote to Colt on the night of April 1 aboard a steamer on the Ohio River bound for New Orleans. Walker and his men were given a grand sendoff with women waving handkerchiefs until the steamer was out of sight. Three more weeks, and still no revolvers. After another week or so rounding up more recruits, Walker expected to leave for Vera Cruz. On April 28 he penned a last letter to Colt from New Orleans: "I have nothing more to write than merely to re-express my hope that you will make some arrangement to forward my Pistols direct to Vera Cruz. . . ." The Colts did come to Vera Cruz eventually but ended up stalled in a depot. When Walker finally rejoined the war in May, the guns he had helped Colt design languished unused.

With June approaching, Daniel Wesson and Cynthia Hawes decided that they had waited long enough for her father's approval. Cynthia would soon turn twenty-two; Daniel was already there. Their relationship was solid. Time to take a bold step: elope. Thirty-odd miles from Northborough across the Massachusetts-Connecticut border was a town perfectly suited to the purpose. Thompson was a well-known quickie-wedding destination for New Englanders. Neighboring states insisted that couples publish their intentions to marry two or three weeks in advance. Not so in Connecticut, where the only requirement was a single brief notice from the pulpit. Since Thompson was close to both Massachusetts and Rhode Island, couples from those states could ride there on a Sunday morning, which they often did, and be promptly married at Stiles Tavern, where the sociable landlord and Justice of the Peace Captain Vernon Stiles (his barroom was also the local Democratic Party headquarters) performed the services "with a grace and sympathy that charmed all participants. Scarce a Sabbath passed without bringing wedding parties to partake of the frosted loaf always made ready for them." For their wedding, however, Daniel and Cynthia opted for the Reverend Doctor Daniel Dow, the seventy-five-year-old, long-serving pastor of the local Congregational Church, to unite them, which he did on May 26, 1847. Whether he liked it or not, Luther Hawes now had a son-in-law.

The day before the Wessons began married life, Samuel Walker rode into Perote Castle, the same place where fellow Texans had been prisoners four years earlier and which the Americans had captured a month before. For the Texans, Perote had become a symbol of suffering, a place to inspire retribution. Having picked a white bean in 1844 while seventeen of his fellow Texans picked black and were executed, Walker stoked his own retributive fire. If Santa Anna had had his way then, Walker knew, he too would have been shot. Now he was part of a conquering army. A time for personal vengeance had come.

10

WALKER GETS HIS COLTS

Our banners floating, on we'll go,
To conquer all of Mexico.

A. M. WRIGHT, *A SONG FOR THE ARMY*, 1846

Within three months of arriving in Vera Cruz, Walker wound up confined in gloomy, moated, thick-walled Perote Castle. This time, though, it wasn't the Mexicans who locked him up; it was his own compatriots. Walker had accused his superior of cowardice during a skirmish and was confined to quarters for insubordination. But General Joseph Lane turned Walker loose. If the Americans were going to fight Santa Anna and wrap up this dirty and increasingly unpopular war, the charismatic, hard-fighting Walker needed to be in command of his own company.

It was now the first week of October 1847. Three weeks earlier, Mexico City had fallen to the Americans, but still the war continued, with bandits and guerrillas as well as Mexican regulars harassing US forces. Guerrilla warfare was a Texas Rangers specialty, and while Walker was now a captain in the Army, he still thought and acted like a Ranger. So did his men, which was fine with the Army as long as they maintained proper respect for property and noncombatants, which they often did not.

For that reason Mexicans in the area around Perote feared and loathed Walker's horsemen.

The enmity was reciprocated. "Should Capt. Walker come across the guerillas God help them, for he seldom brings in prisoners," one fellow soldier wrote in his journal the previous June. "The Captain and most all of his men are very prejudiced and embittered against every guerilla in the country." Another diarist observed that Walker had "an inveterate hatred against the Mexicans and when he has the power he carries on the war according to his own peculiar feelings."

Walker was missing a certain power, though: Colt's new pistols. He knew they were in Vera Cruz, but he couldn't get permission to retrieve them. If he had the pistols, Walker thought, he would surely overwhelm Mexican forces wherever they were and capture or kill the general, referred to by Americans as "the Napoleon of the South." Walker dreamed of enabling the annexation of Mexico, to "open a new and extensive field for the display of American genius and enterprise," as he described it in a letter to his brother. He had his gorgeous Wesson rifle—and still owed Edwin a final payment, which he told his brother he would send with a soldier who was on his way home—but its craftsmanship surpassed its strategic value. His compatriots admired the rifle, but it had only one shot. Colt's pistol could deliver six in rapid succession.

Then fate favored Samuel Walker at last. On October 5 a pair of Colt's new revolvers arrived, not as part of the shipment stalled in Vera Cruz, but as a personal gift from the inventor, who told Walker in a letter written more than two months earlier that he hoped "they will prove substantial friends in time of need." The long-desired invention of Colt's and Walker's collaboration had made its way from New Haven to New Orleans then to Vera Cruz aboard the steamer *Martha Washington,* then inland, and finally into Walker's hands. Officers spoke "in the highest terms of them and all of the Cavalry officers are determined to get them if possible," Walker wrote to his brother the night the guns arrived. "Col. Harney says they are the best arm in the world. They are as effective as the common Rifle [at] one hundred yards and superior to a musket even at two hundred yards."

Walker finally had something to put into his saddle holsters that was infinitely better than the obsolete single-shot pistols the Ordnance Department had doled out to the dragoons. If his troops didn't have their own Colts, at least his mammoth six-shooters were with him at last, as ready for action as he was. And they were just what he had ordered, down to the blade-shaped front sight fashioned of German silver. Colt had even added a touch of artistry, something he had started with his Paterson revolvers. Around each cylinder was a scene created by master New York engraver Waterman Ormsby, inventor of a roller-die cutting process to embed art on curved surfaces. At Colt's insistence Ormsby's scene was a tribute to the Texas Rangers, depicting a romanticized version of the Battle of Walker's Creek with the Rangers dressed in military uniforms, not the frontier garb they actually wore. Both Jack Hays and Samuel Walker were in the foreground—Hays galloping on a light horse, Walker on a dark one—each man holding a revolver as they chased Comanches in full flight. The barrel of Walker's just-fired gun poured smoke, while the Indian close ahead tumbled backward off his horse.

Four days after his Colts arrived, Walker was bound for glory. He and his troops were part of General Lane's column of three thousand men on their way to Mexico City. As they approached Huamantla, a town in the central Mexican highlands, Santa Anna was waiting for them, hoping that Lane would pass by and expose the rear of his column to a surprise attack. But Lane had heard the Mexican general might be in Huamantla, so that night he sent spies into the town to find out. Santa Anna and most of his cavalry had left the day before, the spies reported, but some artillery was still there. Lane decided that his mounted troops, including Walker and his men, should ready themselves to march a dozen miles to Huamantla and take the town.

Early on the morning of October 9, the rattle of infantry drums and the sounds of cavalry preparing for battle roused Lane's column. Rain had been heavy in recent days, but this morning dawned bright after a starry night, the sun's rays "gilding the peaks of the distant mountains, the church spires in the neighboring villages, the white-walled haciendas and dark groves upon the hill-sides, till all resembled a sheen of gold." The ground, damp beneath the soldiers as they had slept, dried quickly. After roll call the men breakfasted on hard crackers washed down by coffee and then gathered in formation to await orders. By now a rumor had begun to circulate: Santa Anna was indeed there. Everyone was delighted. Maybe they could capture him.

By eleven o'clock, more than a thousand Americans were on their way to Huamantla, warm breezes rustling their banners, their cavalry's horses kicking up clouds of dust from now dry terrain. The mounted men, including Walker and his company, were ordered to keep some distance in advance of the main body. Five miles out of town, Walker decided to push ahead even faster. He and his two hundred riders broke from the rest at a quickened pace, hoping to catch the Mexicans by surprise. A narrow alley flanked by sharp-leafed maguey plants leading into Huamantla forced the men to trot two abreast until they reached the center of town, where Walker ordered them to draw sabers and charge. With a wild yell the riders spurred their horses into a gallop, hooves clattering over paved streets, and went at the Mexicans, who were turning their artillery toward them and firing when they could. With a slash of his blade, Walker decapitated a man trying to light a cannon's fuse. The five hundred Mexican lancers still in Huamantla fled from the Americans' furious assault, Walker's men pursuing, cutting down some as they tried to find safety. Walker went through town to the outskirts, searching for Mexican artillery that had been left behind. Then his company began securing the cannons they had captured.

What they did not capture, to their dismay, was Santa Anna. By the time the troopers attacked, he had left Huamantla with the bulk of his cavalry, as Lane's spies had reported. But he wasn't far away. From a hill within sight of the town, Santa Anna saw what had happened. He ordered his lancers into action.

The American infantry was a mile and a half from Huamantla when they heard gunfire. Walker, they knew, had just engaged the enemy. Then they saw the Mexican lancers, two thousand of them, racing toward the town. "They made a most magnificent appearance, dressed as they were in red and green uniforms," remembered one soldier, "and I never beheld a more animating display. Their long and bright lances reflected the rays of the sun, and flashed like a sea of diamonds. The crimson pennons of their lances fluttered gracefully from the staffs, while above the rest was the national flag of Mexico. . . ."

The infantrymen were tired from marching, but they had no choice except to race to Huamantla as fast as they could to rescue Walker's vastly outnumbered force. The Mexican lancers got there first, their horses covered with foam and sweat from hard riding, and hemmed in the Americans by blocking every street. At Walker's order, his men turned captured cannons on the Mexicans, trying to fire into their midst, but they had no fuses to do the job. One man tried unsuccessfully to use his pistol to set off a cannon. In a convent yard by a large house on the corner of the main square, Walker kept shooting and giving orders, when one bullet tore through his chest and another into his head, dropping him to the ground, dead.

Lane's infantry pushed the Mexican soldiers back out of Huamantla and then took charge of the town. They were not gracious in victory. Walker's death hit them hard. Those who knew him wept openly. General Lane, known for his hot temper, turned the soldiers loose, letting them ransack, rape, murder, and plunder as they wished. There were "shouts, screams, reports of fire arms and the crash of timber and glass as the troops batte[red] down the doors and windows," remembered a lieutenant. "Dead horses and men lay about pretty thick, while drunken soldiers, yelling and screeching, were breaking open houses or chasing some poor Mexicans who had abandoned their houses and fled for life." Santa Anna had escaped, but his military career was dead. After the battle of Huamantla, the Mexican government relieved him of command. The main battles of the Mexican-American War were now over. The ugly aftermath of Huamantla, in which Americans savaged their defeated foe, would be an indelible stain on American honor.

Walker became, in death, even more the hero that he'd always wanted to be in life. His posthumous fame imbued Colt's new revolver with renown. It didn't matter that the gun dubbed the "Walker Colt" still had glitches, that it exploded, for example, if loaded with too much gunpowder or that its ramrod lever would sometimes fall down and block the cylinder from turning after the revolver was fired, thereby making a follow-up shot impossible until the lever was shoved back into position. A war hero had led the successful invasion of a Mexican town with Colts on his saddle, and the US government wanted more of them.

Walker was dead, and Colt's dream was fully, finally alive.

11

THE FALL OF EDWIN WESSON

Samuel Walker died with a Colt revolver in his hand, not a Wesson rifle, and this was a problem for Edwin. With his most famous patron gone, the chances were now slim that he could land a contract with the Army. But Edwin didn't give up. Instead he kept thinking big and aiming high. He decided to move forward with a plan to relocate the family business from its hub in Northborough, Massachusetts, to bustling Hartford, Connecticut, seventy-five miles to the southwest. There the Wessons could spread out into a bigger headquarters.

The move made sense. Although growing in population, Northborough was still a minor town in the 1840s. It was mostly agricultural, with a smattering of industry: mills and textile firms, brickmakers, and boot cobblers. Making combs—ornamental versions as well as ones for grooming—mostly from animal horns and hooves provided by local slaughterhouses had begun there in 1839 and was now becoming the town's largest enterprise. And, of course, Northborough had the Wesson rifle business. What it did not have at the time were speedy transportation routes to larger markets or the kind of waterpower needed to operate large factories. While the lovely Assabet River and the town's many streams were fine for the Northborough businesses, their power paled in comparison to that of the steady, strong Connecticut River making its way through Hartford, whose population was nine times as large. Edwin Wesson wanted power and magnitude, and for it he was willing to abandon his solid reputation as a Northborough citizen. (He was one of five men chosen to decide whether to buy a fire engine for the town.) He and brother Daniel decided to depart for Hartford, taking several workmen and their families with them.

At his rented shop and forge in Northborough, Edwin had managed to produce fewer than 150 guns a year, including target and hunting rifles, the occasional pistol, and relatively small shoulder-mounted guns called "buggy rifles," which often came with detachable stocks. If he was to grab a larger market, he needed to expand the business dramatically. For his Hartford factory, Edwin decided not to rent or buy. This time he would build.

Edwin Wesson chose a plot of land that was 50 by 132 feet at the corner of two downtown streets and close to the freight depot and the Connecticut River, a perfect spot for a would-be titan. Over the winter, as 1847 gave way to 1848, his vision quickly became reality brick by brick by brick, three stories high, covered in metal, ready to make rifles in greater numbers than Edwin had before. He also erected a blacksmith's shop, a coal house, and several other small buildings on his lot. Before the end of summer 1848, most of Wesson's rifle-making machinery was in place, soon to be followed by a locally made, eight-horsepower steam engine to run it.

All of this took lots of money, which Edwin didn't have. A group of local lenders saw him as a good prospect for investment, so in April 1848 they funneled him $4,000 through a loan with Hartford's Society for Savings, the state's first municipal savings bank, which had already lent Wesson $5,500. These lenders had great faith in him, and why wouldn't they, given the press Wesson was receiving? That very month *Scientific American* declared that Wesson was the "most famous manufacturer of rifles in the world." The next month Wesson borrowed another three hundred dollars from George S. Lincoln & Co., whose Phoenix Iron Works made a variety of products, including drills and lathes.

Edwin Wesson thought he had every reason to be optimistic. The previous June he had patented a new gun with seven barrels linked together, all of which fired simultaneously. Wouldn't the military want such a walloping weapon? A writer at *The Hartford Daily Courant* thought so. "This is beyond all doubt the most destructive weapon ever invented," he wrote, and an army that had both Wesson's and Colt's guns "could easily cut up almost any force of an enemy." The writer recommended that the Army obtain them "in large numbers. The Government cannot overlook such an important invention as this." Edwin also heard that a wealthy British engineer and gun fancier who had come to the United States a few years earlier was about to publish a book praising Wesson rifles specifically. The future looked bright for Edwin.

His younger brother's star was also rising. Daniel's indentured servitude was coming to an end, and his responsibilities at the factory were increasing. He was in charge of its mechanical operations, which was a natural role for a man whose penchant for tinkering with machinery had accompanied him since childhood. And he was inventing. In the factory's basement was a machine Daniel designed to plane wooden boards using a heavy railway carriage that could hold raw timber while a wheel with cutters trimmed the wood to uniform width. When the machine completed its task, the surface of the plank would be level, smooth, and straight. Daniel had more responsibilities than just making the Wesson factory run well. He was now a family man, for on May 21, 1848, three days after her husband turned twenty-three, Cynthia Wesson gave birth to a daughter they named Sarah Jeannette.

With Daniel's talent and Edwin's vision, success seemed certain, if you could overlook the debts.

A half mile from the Wesson works, down the Connecticut River, Samuel Colt had started on his next gun. In a rented brick building one story taller than Wesson's, Colt was making a slightly scaled-down version of the Walker revolver that corrected most of the original's bugs. For the first time since his Paterson venture went bankrupt, Colt had a factory of his own. What's more, he and his business were back home in Hartford with a government contract to make two thousand big "Dragoon" revolvers, still powerful but less unwieldy than their predecessors, for use by mounted troops. The secretary of war wanted fifteen hundred more of them in early 1848, but Colt's price was too steep. An early tradition had been that gunmakers and the government would cooperate to a degree, paying less attention to patent protection. But Colt held on to his patent rights. This trendsetting precedent annoyed President Polk, who grumbled that the government couldn't make additional revolvers without Colt's permission, which "the inventor is unwilling to dispose of at a price deemed reasonable."

Edwin Wesson also had patents on his mind, though in his case he was trying to obtain one, not haggle with the Army over it. If his "most destructive weapon ever invented" failed to find a place in the market, Edwin had hopes for a firearm he was trying to patent along with fellow gun inventor Daniel Leavitt. This was a revolver that he said operated differently from Samuel Colt's weapon and was just as worthy of a government contract. Edwin had been submitting applications to the US Patent Office, but each time he was turned down, at least partly because an examiner thought his revolver would infringe on Colt's patent. It would not, Edwin insisted, and he accused the inspector of prejudice, saying the man was in cahoots with Colt. Wesson even visited Washington to plead his case. Despite not having a precious patent, he tried to persuade the Ordnance Department to buy his revolver.

"I know my pistol is not an infringement on Colt," he wrote to Chief of Ordnance Talcott on October 3, 1848. If Colt thinks it is, Edwin continued, "he has only to bring his case into court & let it decide. I have no idea he would ever do this, but as he might, should I make a contract I would prefer that it should be for considerable of an amount." Wesson's request went nowhere.

As far as the extended Wesson family knew, Edwin was doing just fine. His brother Rufus Jr. evidently thought so, because on March 19, 1848, he approached Edwin for yet another loan. "I haven't heard from you since I was at Northboro but I heard you had gone to Hartford," Rufus Jr. wrote. "Our respects to all the folks. We are all well, no news much. I am in want of that money very much in these hard times. Please forward and you will much oblige."

There was every reason for the public to think that Edwin's business was soaring, despite his frustration over failing to patent his collaboration with Leavitt. In early 1849 *The Connecticut Courant* enthusiastically reported that more than twenty workers at the Wesson factory were churning out $30,000 worth of rifles each year, "gaining for our city the credit abroad, of furnishing the most celebrated rifle known in the United States." Daniel was perhaps the only person who knew Edwin's finances were shaky.

On the day he appealed to Colonel Talcott for an Army contract, Edwin's creditors closed in, for the master rifle maker had failed to make good on his loan repayments. On behalf of George S. Lincoln & Co., a justice of the peace ordered the Hartford sheriff to seize $1,200 worth of Edwin's possessions to cover unpaid debts. That same day another creditor filed an attachment on the strength of two promissory notes Edwin had given him, one dated July 26 and the other from the spring before his move to Hartford. The next day, October 4, a local lumber company directed the Hartford sheriff's attention to specific Wesson property. Included in the list were sixteen rifles, three unfinished pistols, thirteen partly finished black walnut gun stocks, two grindstones, a steam engine then in pieces, thirty-five mahogany gun cases, a scale with weights, an office desk whose drawers were filled with assorted tools and gun parts, steel and iron bars, buckets, shovels, a coal house along with all its coal, thirty bushels of charcoal, an anvil, an iron vice and bench, and a blacksmith's shop with all its contents, including a bellows. Everything remained in place while the legal wrangling moved forward. Then there was still the debt Edwin owed on the Society for Savings's loans. For the Wesson operation time and money were running out.

Colt's fortunes kept moving in the opposite direction. By early autumn 1848 he had outgrown his first Hartford factory and moved the business into a new building, this one measuring 50 by 150 feet, not including blacksmith shops, and it was five stories tall in a section of the city filled with new industrial structures. Colt's was a true factory with assembly lines that heralded a new era.

"Now what scene can be more captivating to the eye of a live mechanic," a *Hartford Daily Courant* correspondent asked rhetorically in the first week of 1849, "than those long lines of shafting and machinery, and that vista of busy workmen, in rows the whole length of the building? Each man and each machine intent upon their own business, and knowing just exactly what they are about."

As the winter of 1848–49 approached its end, Colt's factory was in full operation. Edwin Wesson, on the other hand, was mired in debt. He was also in failing health. In late January 1849 he came down with chills and other flu-like symptoms. Chronic attacks of erysipelas, a stubborn bacterial infection of the skin, only intensified Edwin's suffering. Over several days his condition worsened. Infection inflamed his lungs. At 1:00 in the morning of January 31, Edwin Wesson died at his home, leaving a widow, two adopted children, and an indebted dream. He was thirty-seven years old.

In announcing Edwin's death three days later, *The Hartford Weekly Times* told its readers that Daniel Wesson, "a skilful mechanic," was "fully capable of conducting" his late brother's rifle-making business. It was certain, the paper predicted, "that an arrangement will be made in due time, that will ensure the continued manufacture of the much admired 'Patent Muzzle Rifle,' in all its perfection." *Scientific American* echoed that assurance for a national audience: "His brother Mr. Daniel Wesson, takes his place and the manufacture of the rifle will still be continued at Hartford with the same perfection as before." Only ignorance of Edwin's precarious finances could permit the buoyant optimism shown by the *Times,* which told its readers, "had Divine Providence permitted him to live, he would have been one of the wealthiest and most beloved among us."

The bondsmen who had engineered the loans from Society for Savings weren't so sure, at least about Edwin's prospects for wealth had he lived. On the day *The Hartford Weekly Times* published his obituary, they met to decide what to do with his debts. Selling the factory, its contents, and the land would not cover the amounts owed. As for the celebrated patent muzzle that Alvan Clark had invented and then licensed to Edwin, some people thought it was valuable, while others said it would be hard to find a buyer. Besides, Edwin had bought the patent for eight annual installments at a hundred dollars each but had fallen down on that obligation as well; he'd paid Clark only two of them. The creditors would see if Thomas Warner—who had been a master armorer at the Springfield Armory, had supervised the production of Walker Colts at Eli Whitney's armory, and had worked with Edwin Wesson—could partner with Daniel to keep the business going, but they knew that the excellence of Wesson rifles was both a plus and a minus.

"These guns are almost universally admitted to be the most perfect and effective fire-arm ever tested," the bondsmen concluded in a report on their meeting, "but they are at the same time, very expensive, and beyond the means of many persons who would be purchasers at half the cost of this gun." Perhaps a joint stock company could be formed to pay Warner to make the guns and leave finding a market for them to the company. In any event they decided that the Society for Savings should foreclose on the mortgage and take possession of everything.

Daniel seems to have had no financial stake in his brother's business. His contribution was talent and time. But he wanted his tools back. He had used them at the factory, Daniel admitted, but he insisted that they belonged to him and were never Edwin's. He even went to court to argue for their return. A judge disagreed. Those tools, he ruled, were part of Edwin's estate and therefore belonged to his creditors.

A couple of months later, in the late summer of 1849, a minor miracle happened. The Wesson & Leavitt revolver that Edwin had fought so hard for was finally awarded a patent. Now whoever owned that patent could challenge Samuel Colt in the marketplace for a share of the profitable revolver business. Edwin Wesson would be vindicated. In the meantime, though, the Wesson rifle factory—and everything in it—was set for auction at 10:00 AM, November 22, on the land Edwin had bought only two years earlier.

That fall Daniel Wesson closed his brother's now useless order book, the one Edwin had opened with a joke about having no cash, but before he did, Daniel wrote in it something that had nothing to do with facts and figures, or guns that would never be made. In strong black ink, the young gunsmith penned a personal reflection on what had been as well as an admonition for the future:

“Thus ended the manufacture of Rifles by the far famed E Wesson. No thing of importance will come without effort.”

12

THE WRATH OF SAMUEL COLT

After Edwin died, Daniel Wesson stayed on in Hartford, where he started a company to make rifles under his own name. "Wesson" meant something in the world of fine guns, and Daniel had the skill to produce firearms worthy of his late brother—and of himself. What he didn't have then was cash to get going, so Cynthia bankrolled him with four hundred dollars of savings her family had given her.

Daniel had other prospects for making money from guns. Although Edwin Wesson was dead, his latest patent wasn't. His younger brother had high hopes of restoring the family name as the premier brand for elite firearms, either by himself or even better with a mass-produced gun, and the Wesson & Leavitt revolver was the ideal product to realize Edwin's dreams posthumously. After the auctioneer's hammer came down for the last time on Edwin's assets, the moment was near to start making guns again, provided investors could be found to resurrect the business. Going forward, though, would trigger a battle royale in federal court between two of the biggest names in guns.

On March 5, 1850, Daniel Wesson and several other men joined forces to launch a new gun firm, the Massachusetts Arms Company, in Chicopee Falls, an industrial town just north of Springfield near the east bank of the Connecticut River. Among the stakeholders who saw a future in the company was another Massachusetts gunmaker interested in repeating firearms, a man Wesson had met while both worked in Worcester: Horace Smith. Their names would be forever linked. The sole purpose of the new company was to make the Wesson & Leavitt revolver, an ungainly looking handgun, almost fifteen inches long, its cylinder and barrel thrusting far out ahead of its handle as if stretching toward a target. Colt's Dragoons were front-heavy, too, but they had a more balanced appearance, perhaps because the American eye was beginning to get used to seeing them in shooters' hands. The hammer on the Wesson & Leavitt was screwed onto the gun's right side, curving up and forward to the middle on top, while a Colt hammer sat mostly hidden, directly in the middle of the gun. One revolver would never be confused with the other, but some internal workings were similar—perhaps too similar.

Metal on the first Wesson & Leavitt revolvers that emerged from the factory in early 1851 bristled with several patent designations, one as recent as the previous November 26. The company dubbed them "army" revolvers, apparently hoping for military contracts. The marking on a bevel gear, which helped the cylinder turn as the hammer was cocked, would have particularly pleased Edwin: "WESSON'S / PATENT / AUG. 28, 1849." That near-simultaneous coordination of split-second movements—hammer cocked, trigger pulled, cylinder turning, bullet fired—was a major selling point and the key to challenging Samuel Colt in the marketplace. On that much Colt agreed, and he didn't like it.

By now Colt had been making money on several models, including the big Dragoons, a "Baby" Dragoon, and a small pocket pistol for carrying discreetly that was popular with gold seekers out West. He saw himself on track to make even more money with a midsized revolver well suited for sitting comfortably in a belt holster. This was the .36 caliber 1851 Navy, so called because the scene around its cylinder—another Waterman Ormsby creation, like that on the big Walker revolvers—depicted the Texas navy battling Mexicans in 1843. This was Colt's tribute to what was then the Lone Star Republic for favoring his Paterson revolver during his early business struggles. The Colt factory in Hartford was ramping up production with hopes of arming as many people as possible with multi-shot weapons that were superior to the old-style pepperbox. There were customers aplenty.

The cylinder of each Colt turned with graceful precision as the gun was cocked, lining up a loaded chamber with the barrel, readying the weapon for shooting in a single, smooth movement of one hand. Colt insisted that the right to make a gun with such a mechanism was his alone, and he held a patent to prove it, one that he had convinced the Patent Office to extend beyond its original expiration date despite objections from others, including Edwin Wesson. If a well-positioned upstart like the Massachusetts Arms Company could also claim the right to make a gun that performed like a Colt, he would lose an advantage he had enjoyed for years. Everything Colt had built was in jeopardy.

What also rankled Colt was the fact that Edwin Wesson had opposed his efforts to have the government reissue his revolver patent in 1848. Colt was successful then, but now it seemed like the ghost of Edwin Wesson, in the form of this new company, was again trying to defeat him—and with his own former allies. It was bad enough that Thomas Warner had been hired by the administrator of Edwin Wesson's estate to finish more than a hundred rifles that were under construction when Edwin died. Now Warner was making Wesson & Leavitt revolvers for the Massachusetts Arms Company. What's more, Colt had less-than-warm feelings for two creative mechanics who were among the founders of the Massachusetts Arms Company: Joshua Stevens and William Miller. He had hired both away from Eli Whitney in November 1847 to make parts for his Dragoons only to find out the following July that Stevens and Miller were also developing their own revolvers. Infuriated by their disloyalty, Colt fired them. And what did Stevens and Miller do next? They immediately went to work for Edwin Wesson in the very year that Wesson was objecting to Colt's patent reissue.

It was time for Colt to go after the Massachusetts Arms Company with all he had. For that, he needed a fixer like Ned Dickerson.

Edward Nicoll Dickerson was only twenty-seven and had been an attorney for just three years, but he had already made a name for himself in legal circles. Like Colt, Dickerson was passionate about mechanical things, having been inspired by his college friendship with the head of the Smithsonian Institution as well as his schooling in mechanics. Patent law was a natural subject for Dickerson, because it fused his interests: how things worked, who made them work, and the rights and protections afforded to those makers.

Ned Dickerson had another advantage during trials: his looks. When this "strikingly handsome man" rose, his massive, rugged frame—six feet, three inches in height—could dominate a courtroom. "His face, finely moulded, with a keen eye, straight nose and strong mouth, was exceedingly attractive to watch as he talked." And he was an impressive orator, with a touch of showmanship that was indispensable to trial lawyering. Samuel Colt surely appreciated this. Dickerson, with his imposing presence coupled with legal skill, mechanical mind, and gift for entertaining, was just the persuader Colt wanted to take on the Massachusetts Arms Company.

Dickerson wasted no time. He filed suit against the company in federal court in Boston on September 28, 1850. A few weeks later, Daniel Wesson's reliance on the revolver business increased, when he pulled the plug on his solo venture, having made fewer than a hundred rifles. Confident in the righteousness of its own patent, the Massachusetts Arms Company was not deterred by Colt's lawsuit. The firm had been gearing up production, which began in earnest before the end of the year. Both sides knew that the stakes were high.

The Massachusetts Arms Company hired its own super lawyer in former US senator and congressman Rufus Choate, an adept trial attorney whom one newspaper called a "great galvanic battery of human oratory." Four years earlier, in a lurid trial that titillated the press, Choate had defended a man accused of slitting his mistress's throat and then setting fires to burn down the house before skipping town. The killer, he argued, couldn't be held responsible, because the man had been sleepwalking at the time. In his final words to the jurors, trying to spare his client a trip to the gallows, Choate said with a low voice, "In old Rome, it was always practice to bestow a civic wreath on him who saved a citizen's life; a wreath to which all the laurels of Caesar were but weeds. Do your duty today, and you may earn that wreath." The grandiose closing worked. It took the jury two hours to acquit the man.

Now it was Dickerson against Choate, Colt fighting Wesson, in a clash of gun barons that would test the meaning of invention and decide the right to profit from it.

Two teams of lawyers squared off in a Boston courtroom on the last day of June 1851, as the trial of *Samuel Colt v. The Massachusetts Arms Company* began. In his opening statement, Ned Dickerson set the tone: "Some man, who has more brains than discretion, sets himself to work to make an improvement. He is an enthusiast; a poet in wood and steel, as he has been defined. He follows that idea with a degree of enthusiasm utterly inexplicable, particularly when he knows how he will be rewarded for all the time and trouble he bestows upon it." Dickerson spoke of Colt as a lone, struggling inventor, seeking to make something of himself and his creation, toiling for years in service to his dreams, as he fights "to overcome the prejudices of those who are using old contrivances, to beat down the opposition of those interested against him, to bring his invention into use, to make it practical and valuable." Then, when success is near, when the inventor sees the possibility that he will finally reap at least some reward for all he has done, "the infringer steps in between him and the prize."

Dickerson's opening was eloquent but incomplete. He didn't mention Colt's family connections that had helped launch the Paterson gun firm. Nor did he say anything about the personal encouragement Colt had received from Henry L. Ellsworth, a prominent Connecticuter who was to become the country's first commissioner of patents and whose father had been a framer of the US Constitution, a US senator, and the country's third chief justice. In an early 1832 letter from Washington, Ellsworth reported to Colt's father, an Ellsworth friend, that "Samuel is now here getting along very well with his new invention. Scientific men & the great folks speak highly of the thing. I hope he will be rewarded well for his labors. I shall be happy to aid him." Ellsworth did just that over the next few years, introducing Colt to useful people. And Dickerson did not mention Rufus

Porter, a painter, prolific inventor, and founder of *Scientific American* magazine, who, only three months after Colt received his first patent in 1836, sold to Colt for one hundred dollars the rights to a revolver Porter himself had invented; that would have bolstered the defense claim that others had been inventing revolvers like Colt's. Instead Dickerson told the jury that when his client took out his patent, he was "a young man without means, and without many friends. He was a New-England boy. He came from Ware, in Massachusetts, and had made this invention without money to put it into practical operation."

Dickerson made the argument that inventing was not meant solely to enrich the inventor, but to advance the public good. "It is the interest of the public to have the best arm that can be made," Dickerson said. "[M]en who go to tame the untamed forest, and encounter the Camanche [*sic*] or Blackfoot, want something upon which they can depend. The best arm is what the interest of the public requires. They should have one which will do its work most certainly, rapidly, and destructively."

Lawyers for the Massachusetts Arms Company didn't disagree that the public needed the best available gun. Their client's revolver was a fine one, too, they claimed; it had a valid patent and shouldn't be barred from the marketplace. Defense attorney Reuben A. Chapman started to dismantle Colt's mythology. He was no genius pauper who had struggled in solitude only to see his success siphoned away. When Colt got his first patent in 1836, Chapman told the jurors, he was "never a mechanic; never, I believe, having given time to mechanical study; brought up a gentleman's son; having travelled in Europe, and been to India. . . ." He "enlisted many persons of capital—I suppose, some of the very smart young men of New York city, who had plenty of money—in his enterprise, at Paterson. . . ." And now, Chapman said, "Mr. Colt has come here, he has made his threats that he has $50,000 to expend against us, and if we beat him, that he will go into the market and bid us down."

"Have we made such threats?" interrupted George T. Curtis, one of Colt's attorneys. "I have heard nothing of the kind in this cause."

"As a matter of fact it is so," Chapman insisted.

"It is not to your knowledge," Dickerson rejoined.

"He has a right to come here," Chapman said, "but we ask justice."

Later in the trial, Dickerson returned to Chapman's allegation:

> My Brother Chapman alleged, in his opening that Mr. Colt had threatened these parties to spend $50,000 in prosecuting them. I can only say to this, that Mr. Colt would feel delighted if any one could prove, to his entire satisfaction, that he had that much money to spend for any purpose whatever; and, therefore, I suppose the threat was not made. But I will say this, for the satisfaction of our friends on the other side, that whatever he has, will be devoted, if necessary, to defending his rights, whether it is $50,000 or $10,000.

Dickerson brought guns to court as well as patent drawings to show jurors how they worked. He even put on the stand the venerable Thomas Blanchard, the renowned inventor whose many creations at the Springfield Armory, in addition to the lathe that duplicated gun stocks by mechanically copying a model, included a machine that streamlined the making of gun barrels and a steam-powered horseless carriage. His words carried great weight in the world of mechanical creation.

Dickerson asked his expert whether the Colt and the Wesson & Leavitt operated on the same principle.

"They are substantially the same," Blanchard testified.

The defense had two lines of attack. One was to insist that Blanchard was wrong, that the Wesson & Leavitt patent was sufficiently different from Colt's. But it was becoming increasingly clear that the two mechanisms were distressingly alike—distressingly from the Massachusetts Arms Company's point of view, that is.

The other line of attack was more promising: to show that Colt was not the first inventor to make a revolver with a cylinder that turned and locked when its hammer was pulled back. If others had made the breakthrough before him, Colt would have no claim to originality. Rufus Choate brought more than one old revolver to court as proof that the concept predated Colt's success. As the evidence was presented, it appeared that perhaps Choate had a point. Even Colt attorney George Curtis worried during the trial that his side could lose. He and Dickerson would not yield the point, of course. It was the jury's job to decide who was right. Then, thanks to Dickerson's extensive trial preparation, liberally funded by his client, something happened that threatened irreparable damage to the Massachusetts Arms Company's case.

Dickerson managed to pry from a defense witness a startling admission. A part in one of the supposedly old revolver mechanisms that defense attorneys said were similar to Colt's was not original and had been made to look as old as the rest of the gun. Was that an attempt to defraud the court, to put one over on the honest men of the jury? Dickerson said as much, making sure the jurors got the point.

Testimony ended at the beginning of August. It was time for the attorneys to give the jury their closing arguments, scheduled to start on Monday, August 4. When the trial had begun at the end of June, it was Reuben Chapman who gave the opening statement for the Massachusetts Arms Company, while the all-important closing was to be left to the eloquent Rufus Choate, who had limited himself to questioning witnesses. But in a blow to the defense, Choate was sick in bed when his moment arrived. The trial was delayed a day in the hopes that he would recover, but on Tuesday, Choate was still confined to his bed. Chapman took over, apologizing to the jury that he hadn't had time to prepare properly.

Nevertheless, Chapman did the job with thoroughness. After walking one-by-one through the witnesses' testimony, he told the jury, "Take, then, our pistol and take his. The great differences between the two are obvious to the eye." What's more, Chapman argued, Colt had rushed through his patent extension, which was actually a major overhaul of the mechanism, in violation of established procedure, for the sole purpose of derailing Edwin Wesson's patent application. This, Chapman insisted, was a kind of fraud perpetrated by Colt on another inventor.

George Curtis then spoke to the jury on Colt's behalf, returning to the theme with which Dickerson had opened the trial: the individual inventor against rapacious corporate parasites.

> [T]here are now living in this country as many persons of inventive genius, whose inventions have produced a striking, important and most beneficial effect on the civilization of the age, as in any country in Christendom. Yet these persons, as soon as they have sought the protection which the Constitution and laws of their country undertake to give them, as soon as they begin to reap the fruits of what may be called success, must enter on that field of strife and litigation, of unfair and unjust competition, which the eagerness of rivals, imitators and copyists spreads everywhere around.

After the lawyers had their say, it was up to Judge Levi Woodbury to instruct the jurors on the law. Woodbury was no run-of-the-mill jurist. He had been a New Hampshire state judge, secretary of the navy and of the treasury under Andrew Jackson, governor of New Hampshire, and a United States senator. A "broadly cultured lawyer," Woodbury lectured on a variety of topics while sitting on the Supreme Court, such as "Traits of American Character" and "The Right and Duty of Forming Independent Individual Opinions." He had also run for the Democratic nomination for president. In Woodbury's view America's mission was to overtake other countries in commerce and technology and show "that our new theory of private rights and public duties is conducive to progress in every thing useful. . . ." Now he was a Polk-appointed United States Supreme Court justice doing trial duty. And before him was a case involving private rights and technological progress. He was in his element.

"I recommend you, gentlemen," Woodbury told the jurors, "to commence the investigation of this controversy . . . with a feeling of no hostility or prejudice against [the Massachusetts Arms Company], because they happen to be a corporation, or happen to be a probable overmatch for any single individual." Do only what is legally required, he said; don't let prejudice sway you. But his instructions leaned one way. "[Y]ou should not let your sympathies go beyond the rule of law and duty, because [Samuel Colt] stands alone, and because he has evidently been struggling for fifteen or twenty years on this subject, to do something which might confer a benefit upon his country and reward his own exertions."

When Woodbury was done, the jurors went off to deliberate. It didn't take long for them to come to a conclusion: The Massachusetts Arms Company had violated Samuel Colt's revolver patent.

By the time the jurors made their decision, the Massachusetts Arms Company had produced about eight hundred Wesson & Leavitt revolvers in different sizes. There would be no more, at least none that used the Colt cylinder system as long as Colt's patent lasted. Although the company took its case higher up the legal ladder, the judgment against it would stand. And the message would go out to other firearms makers: Infringe on Samuel Colt's patent at your peril.

Neither Horace Smith nor Daniel Wesson had testified at the trial. There was no reason for them to, since they were only investors. And they hadn't bet their future on the Massachusetts Arms Company crushing Samuel Colt. Not all of it, anyway. As usual Smith had mechanical ideas of his own, including modifications to Walter Hunt's Volition Repeater, which had already been improved by a gunsmith named Lewis Jennings. But the thing still wasn't right. Smith's contribution, if he succeeded in getting it patented, would simplify the mechanism and smooth its operation. Maybe then the brilliant but flawed concept that had sprung from Hunt's hyperactive mind could be on its way to becoming something that might really function.

On August 26, nearly three weeks after the Massachusetts jury handed Samuel Colt his victory, the United States government gave Smith the patent he wanted. At the age of forty-two, Smith had spent many years with guns—making them, refining them, figuring out how to improve them—but this was his first patent. It moved the technology of rapid-fire weaponry one step closer toward reality. But he didn't really hold on to the right. Instead he assigned his patent to merchant and venture capitalist Courtlandt

Palmer, who had been spending his money backing efforts to make a profitable repeating firearm, so far without success. He had hired Smith to reverse his misfortune with guns. In less than three years Smith would have a second repeating-gun patent, one he would share with Daniel Wesson; this would be Wesson's first patent. Both men were now patent-war veterans. They would grip their new rights tightly—for a year.

13

CONQUERING THE EVIL COLLAR

The first two days of February 1848 brought watershed events: one for the country, the other for Oliver Winchester.

On February 2 the Treaty of Guadalupe Hidalgo ended the Mexican-American War. Despite its conciliatory title—Treaty of Peace, Friendship, Limits, and Settlement between the United States of America and the United Mexican States—the document meant that Mexico had lost the war. More than that, Mexico had lost a huge chunk of its territory (more than half, if you include Texas, whose independence Mexico had never recognized), though the country did receive a $15 million payment from the United States, which also agreed to assume more than $3 million in debts Mexico owed American citizens. Mexico's northern neighbor had thus become the transcontinental nation President Polk wanted it to be. The land the United States of America controlled now began at one ocean to the east and ended at another ocean nearly three thousand miles to the west. National pride swelled. Many saw triumph as the natural course of events for a country blessed by Divine Providence and cultural—many thought racial—superiority made manifest in the progress of civilization furthered by industrial excellence.

Not every American was happy with the war's result, though. Many Democrats wanted their country to take over all of Mexico, while Whigs, who thought waging the war was morally wrong from the start, wanted none of the territory American soldiers had conquered. But the matter was now settled, and the Treaty of Guadalupe Hidalgo would eventually receive the Senate's blessing.

The other watershed event took place on the day before the treaty was signed, when the United States Patent Office awarded Oliver Winchester a patent. It wasn't for a gun or anything remotely dangerous, but for a new way to make shirt collars with a method that Winchester himself had devised. There was an "evil" (Winchester's word) in how suspenders atop the shoulder pulled down on the neckband of a man's shirt, causing discomfort to the wearer. Winchester's "remedy" (also his word) was to change the way the cloth would be cut. Accompanied by drawings made with the delicate precision expected of hopeful patentees, his application showed how this could be done. Simply put, Winchester's shirt would have a curved seam running along the top of the body from the base of the neck to the tip of each shoulder. A standard men's shirt at the time had a straight seam between those two points, and it did not touch the body. Suspenders would pull the shirt cloth down onto the shoulders, dragging the collar with it; not so with Winchester's design. His idea might seem obvious now, but it wasn't back then, given the styles of the day, so Winchester got his patent. But what's a patent without a market?

The years Winchester ran a clothing business taught him something about markets. He knew how men wanted to dress, and he was convinced his patented shirt process would attract customers who would then spread word of its wonderfulness. Baltimore, important town though it was, might have suited someone whose ambitions stopped with local customers or men in need of proper attire who happened to be passing through, but Winchester's appetites were greater. He needed to be elsewhere, someplace where he could plunge into a larger market, even a combative one, with a product no one could legally make or profit from without his say-so and where no one who knew him would doubt that the stocky man with the bulldog face and immovable presence would defend this right against anyone tempted to infringe. When he applied for the patent, Winchester was still in Baltimore. By the time it was granted, he was orchestrating his return to New England with Jane and their two children, putting down roots not in hometown

Boston, but in New Haven in time for Jane, a month shy of her fortieth birthday, to give birth to daughter Hannah Jane, whom the Winchesters would call Jennie.

New Haven was more than the city hosting Eli Whitney's armory, the gunmaking enterprise that had saved Samuel Colt from obscurity by producing the inventor's Walker revolvers. Once a leading center of trade in Connecticut, New Haven was now home to a bustling community of many new industries, a place known especially for carriages and clocks, operations that drew a workforce wise in the ways of manipulating woods and metals into finer things that found their own markets. The year before the Winchesters resettled there, an innovative clockmaker named Chauncey Jerome completed moving his business from Bristol thirty miles south to New Haven, where he built timepieces with mass-production techniques. Clock movements had largely been wooden, but Jerome helped pioneer making them of stamped brass in the same factory that made the clock cases. Metal movements that didn't warp like wood meant that their timekeeping wouldn't be thrown off by changes in climate, and they could endure sea voyages to markets abroad. Within a few years Jerome's clocks and others made in America would reliably tick away on shelves across the Atlantic. And they would also tell time in modest homes throughout America, not just where the wealthy lived.

The Industrial Revolution may have brought clock-buying within the reach of people with limited means, but the New Haven carriage trade had richer people in mind. There had always been wagons and carts, but four-wheeled carriages were for the well-to-do. Thanks partly to the elder Eli Whitney's cotton gin—and to the growth of slavery it made possible—the South was doing well financially. At least plantation owners and those who depended on them were, and carriages had become symbols of prosperity in which the well-heeled could show off on race days or when going to church or anywhere the public gathered. Sometimes people who claimed elite lineage would display family crests on carriage doors, letting those on the outside know the worthiness of those on the inside. Since the 1830s New Haven was widely recognized as a source of first-rate carriages that were sent on ships down the coast, to the Caribbean, and even far away to the Pacific islands. But it wasn't just the carriage and clock businesses that brought prosperity to the city's growing industrial sector by the time Oliver Winchester arrived. A variety of trades meant that New Haven was no one-industry town.

The Baltimore shirtmaker had come to New England to dress the many, not the few, but he started small, carrying cloth under his arm or over his shoulder on his way home, where he would cut and stitch it himself. Jane helped make the shirts. Soon Winchester had enough customers to enlist local women to do the sewing in their homes, a system common at the time. Cutting cloth—the heart of the tailoring trade, which required experience and talent—was still done in his factory. Business was good, so Winchester moved his operations to a bigger house nearby that had been a girls' school run by an Episcopal priest. But Winchester wanted even larger markets, the kind that Chauncey Jerome's timepieces reached. If mass-produced clocks were now for the Everyman, why not clothing?

The moment was perfect. Ready-made clothing for men, but not yet for women, was coming into its own. New York and New England merchants sent off-the-shelf clothes beyond the region, as making apparel at home, so common in rural communities during colonial times, was on its way out. "Indeed," observed a journal of commerce in 1840, "the art of household manufacture is fast being totally lost; and the farmer is becoming quite as dependent for clothes upon the manufacturer, as the manufacturer is dependent for food upon the farmer." Tailors were experimenting with new patterns for clothing based on standard proportions, not individual measurements, another step toward mass production, while more and better transportation networks allowed goods to be sent farther from the source. And the birth in the 1830s of the penny press—cheap, widely circulated newspapers—gave manufacturers a chance to advertise their products to larger readerships.

With a proprietary shirt patent firmly in his grasp and a growing business under his command, Winchester was well-positioned to make men's clothes en masse. But to get the most from the new world of mass production, he could use a partner, preferably an established businessman who knew the game and the product. That man was John May Davies.

For Davies clothing had been a father-son business. When he was two years old, his father, Luke, started making leather and cloth caps, supposedly the first person in the country to do so on a large scale. Based in New York, Luke also had a small store near Yale in New Haven, where he sold shirts to college students, eventually expanding into a variety of men's furnishings and making his son a partner in the business. His reputation growing, Luke won a prize "for the best assortment of cloth and leather caps, and oil silks" in 1829. By 1837 when his father retired, twenty-four-year-old John Davies was in New York City, running one of the largest clothing importing and manufacturing concerns in the country, employing at least two hundred workers, who made a hundred thousand caps to send south each year and many more for markets in the North. And the Davies business kept growing.

The pairing of Oliver Winchester and John Davies was a natural. Rather than compete for market share, they combined forces: Davies with his network and enduring reputation and Winchester with his drive, talent, and especially his patented shirt collar. With a loan from Davies, Winchester's shirt business took off even faster, and Davies began selling the patented shirt, telling the public that his own firm could outfit them in shirts "Of Linen, Cambric with Linen bosoms, and Fancy Cambric, cut on an entire new principal [*sic*] and warranted to fit." The two men were solid partners. More than that, they were becoming friends.

Winchester's Baltimore store remained a lucrative venture. Brother-in-law Cyrus Brett, who had represented Winchester in the estate matter that had brought the twins to Baltimore, had also moved to New Haven, leaving O. F. Winchester & Co. in the care of two partners Oliver had enlisted while he was in New England. Twin brother Samuel, who would live the rest of his life in Baltimore, also sold the Winchester clothing line, eventually bringing two of his sons into the business. By the early 1850s the Baltimore store had relocated to Carroll Hall, an elegant new building with European flair that was perfect for selling fine clothes. A meeting hall on the second floor hosted all sorts of events, including political and labor gatherings, commencements, and balls. Shops, including Winchester's, were on the ground floor, well placed to draw in customers. Something new added luster to what Winchester had to offer in his Baltimore emporium. No longer would he sell just standard men's clothing, but also "Patent Shoulder Seam Shirts, made to order and warranted to fit," just as he was selling in New Haven and beyond. Business in Baltimore was good—his shirt bosoms and collars won a silver medal at the 1851 Maryland Institute fair—so good, in fact, that Winchester appealed for help through the local newspaper: "NOTICE TO THE LADIES.—From 30 to 50 neat SEWERS can find constant employment on application at the Patent Shirt manufactory, No. 145 Baltimore street, under Carroll Hall," read one advertisement in *The Baltimore Sun*; "no others need apply, as none but the best will be employed."

In 1851 guns still weren't on Oliver Winchester's mind, but plants were. That year he became a director of the New Haven County Horticultural Society, thanks to his burgeoning interest in flowers, fruits, and other green life. And, of course, his increasing wealth enriched his standing in the community. But his enthusiasm for growing things was genuine. He brought a bouquet of mixed flowers to the September 3 weekly exhibition and a plate of large pears to the Society's annual fair.

While Winchester was enjoying horticulture and making his patented shirts in New Haven and Baltimore, Samuel Colt was busy forty miles away in Hartford, making guns and enjoying the aftermath of crushing the Massachusetts Arms Company in court. He would have the perfect occasion to show off his wares. Great Britain was about to host a world's fair of industry, where countries could display whatever marvelous things they made, and all would hail the new era of machine wonders. For the British Empire it was an occasion to bask in the glow of its own excellence. For Samuel Colt it was a chance to be a showman once again, this time on a world stage. Ten days after his legal victory in Boston, Colt was aboard ship bound for London.

14

TO LONDON 'MIDST IRON AND GLASS

Here you may make yourself acquainted with the new method of vaccination as performed by the practitioners of the Far West upon the rude tribes who yet incumber the wilderness with their presence.

THE TIMES OF LONDON MAY 27, 1851, DESCRIBING TONGUE-IN-CHEEK A DISPLAY OF COLT REVOLVERS AT THE GREAT EXHIBITION OF THE INDUSTRY OF ALL NATIONS IN LONDON

Samuel Colt set sail for England, fresh from his victory over the Massachusetts Arms Company, on a course for another triumph. The Great Exhibition of the Industry of All Nations—perhaps the first true world's fair—had been under way in London for more than three months. Colt's guns were already there, of course, more than five hundred of them, cared for by an assistant with strict instructions on how to display them and orders that no one was to shoot any until the inventor himself arrived. This gala gathering of industry was just the opportunity Colt craved to shine his light brighter and be seen farther than ever. And the modest American presence needed a showman like Samuel Colt.

The building erected specifically for the exhibition was spectacular, unlike anything seen before in London or, for that matter, Europe at large. It rose over Hyde Park like an Olympian greenhouse (it was designed by a master gardener) with more than a thousand iron columns supporting three hundred thousand panes of glass over eight hundred thousand square feet of floor space. Two hundred miles of crossed iron bars held the structure together. To spare three magnificent elms from being leveled, a barrel-vaulted transept was built over them, glass and iron arcing high over their tops like the celestial reach of a cathedral's ceiling, letting the sun through to continue nourishing them. Future prime minister Benjamin Disraeli called the airy building "an enchanted pile . . . raised for the glory of England, and the delight and instruction of two hemispheres." It was called the Crystal Palace.

The Great Exhibition was conceived partly as a celebration of peaceful industrialization. In that spirit British pacifists lobbied to exclude displays of "such weapons as were constructed only for the destruction of human life." A letter to the religious British newspaper *The Nonconformist* insisted that barring "mechanical life-destroyers" from the exhibition would show "a desire for peace on the part of England, which could not fail to have a beneficial effect" on the world at large. The pacifists lost this battle.

Peace was a laudatory goal, but the Great Exhibition was also an opportunity for the British to show off achievements that put them in the first rank of modern nations, a task they took up with enthusiasm. The middle of the nineteenth century was a time of patents, of designs, of new ways to forge and bend metalwork to serve the good of mankind and the advancement of civilization. And what better place for such an exhibition than the heart of the British Empire, the birthplace of the Industrial Revolution?

The doors of the Crystal Palace opened to visitors on May 1, 1851, a sun-filled day but for a brief late-morning shower. Queen Victoria did the honors. Only thirty-one years old but already almost a decade and a half into her reign, her majesty was dazzling: dressed in pink crinoline with sparkling jewels and shining silver embroidery, perfumed with an unmistakable air of royal command. At the queen's side was her husband, Prince Albert, who was passionate about free trade and inventions and had been a moving force for the Great Exhibition from the beginning. In front of the opening-day crowd, Victoria prayed for God to bless "this undertaking," that it "may conduce to the welfare of my people, and to the common interests of the human race, by encouraging the arts of peace and industry, strengthening the bonds of union among the nations of the earth, and promoting a friendly and honourable rivalry in the useful exercise of those faculties which have been conferred by a beneficent Providence for the good and the happiness of mankind." After the archbishop of Canterbury bestowed his blessing and a choir sang Handel's Hallelujah Chorus, the queen declared the Great Exhibition open, as "a flourish of trumpets" blasted the news to the outside world.

The panoply of industrial and artistic products spread throughout the Crystal Palace was more extensive and varied than anyone could have imagined beforehand. What was believed to be the world's largest mirror ran the length of the building's main avenue, and an envelope-making machine folded forty-five pieces of paper every minute just so, as if its metal protrusions were a secretary's gentle fingers. In addition to marvels of modern industry, fine arts, and raw materials, there were oddities, like the "tempest prognosticator," a living barometer with a dozen leeches housed in small bottles. When a storm approached, so the theory went, the leeches would react to the change in pressure by climbing out of confinement and triggering small hammers that would ring a bell. The more leeches disturbed enough to escape their bottles, the more the bell would ring, indicating a severe storm was on its way. The thing actually worked.

The night of the opening, after she returned to her own palace, Queen Victoria wrote in her journal, "Dearest Albert's name is forever immortalized." It wasn't long before the Great Exhibition was declared a success. Not surprisingly, Britain's offerings triumphed. Crowds poured in to marvel at the nineteen acres of exhibits and the building itself. By the time it closed, more than six million visitors had seen the vast array of art and industry from England and other parts of the world, bathed in the natural light that the glass ceiling allowed to come through.

Wrote Charlotte Brontë, author of *Jane Eyre:*

> It seems as if magic only could have gathered this mass of wealth from all the ends of the earth—as if none but supernatural hands could have arranged it thus, with such a blaze and contrast of colours and marvellous power of effect. The multitude filling the great aisles seems ruled and subdued by some invisible influence. Amongst the thirty thousand souls that peopled it the day I was there, not one loud noise was to be heard, not one irregular movement seen—the living tide rolls on quietly, with a deep hum like the sea heard from the distance.

Not everything about the Great Exhibition pleased Queen Victoria. On May 19 she returned to the Crystal Palace, this time to see what the foreign section had to offer. When she came to the exhibitors from the United States, she found them "certainly not very interesting." Her majesty could hardly be blamed. The Americans had requested considerable space, then had great difficulty trying to fill it. Their offerings were frightfully mundane compared to opulent Victoriana and finery offered by other regions of the world.

The British satirical journal *Punch* delighted in the United States' paltry presentation. The Americans excused their shortcomings, the journal said, by claiming that their industrial inventions were "too gigantic," too profound in nature. "[T]he reality is so impossible to be understood or described," according to *Punch,* "that the only way to give us any idea of it was to leave it all to our imagination."

Perhaps, *Punch* suggested helpfully, the empty space at the American section could be offered as lodging for visitors to the Great Exhibition:

> By packing up the American articles a little closer, by displaying COLT'S revolvers over the soap, and piling up the Cincinnati pickles on the top of the Virginian honey, we shall concentrate all the treasures of American art and manufacture into a very few square feet, and beds may be made up to accommodate several

> hundreds in the space claimed for, but not one-quarter filled by, the products of United States industry.

The Colt revolver, though, was getting favorable attention even before its ebullient salesman arrived. The Duke of Wellington, hero of the Battle of Waterloo, who turned eighty-two the day the Great Exhibition opened, was also impressed by Colt's work. Several times he returned to the display for another look and was heard praising the value of repeating guns. "The most popular and famous invention of American industry," *The Times* of London told its readers in early June, "is a pistol which will kill eight times as quick as the weapon formerly in use."

To entice sales from countries with colonial dominions, a notice attached to Colt's revolver display quoted from an American military report that "on the Texan frontier, and on the several routes to California, the Indian Tribes are renewing their murderous warfare, and a general Indian war is likely to ensue, unless bodies of mounted men, efficiently equipped for such service, are employed against them. . . . A few bold men, well skilled in the use of these weapons, can, under such circumstances, encounter and scatter almost any number of savages."

The British military took note. That summer they were fighting yet another war in South Africa, where Xhosa warriors had been decimating their soldiers. *The Maidstone Gazette*, a newspaper in a town that was the British military's chief cavalry depot in Britain, observed:

> Amongst the most humiliating circumstances in the recent war at the Cape, was that of a horde of Kaffirs rushing on a small detachment of our troops, and wresting their muskets from their hands after the first discharge, before they could reload. . . . The musket may be an admirable weapon for operating on close masses of men, but in the irregular impetuous rush of such warlike tribes as the Kaffirs, the Affghans, the American Indians, and the New Zealanders, a different description of weapon is requisite, which will

> give the largest number of shots at a given time. Had the Kaffirs been attacked by cavalry, armed with such weapons . . . the enemy would by this time have been hunted out of the colony.

An American item the British public liked was *The Greek Slave,* a marble statue of a young woman, nude with her hands in chains and holding a small cross, kidnapped by Turks to be sold for sexual pleasure. She "stands exposed to the gaze of the people she abhors, and awaits her fate with intense anxiety, tempered indeed by the support of her reliance upon the goodness of God," explained her sculptor. The irony that this lamentation about slavery came from a country that continued to profit from people in bondage was not lost on the British. Again, from *Punch:* "Why not have sent us some choice specimens of slaves? We have the Greek Captive in dead stone—why not the Virginian slave in living ebony?"

Despite the country's continuing tolerance of slavery, at least in the South, the United States' offerings at the exhibition were fruits of a democratic republicanism evident even in the catalogue. Examples of American industry tended to be different from those of other countries. Noted a catalogue of the exhibition:

> The expenditure of months or years of labour upon a single article, not to increase its intrinsic value, but solely to augment its cost or its estimation as an object of virtû, is not common in the United States. On the contrary, both manual and mechanical labour are applied with direct reference to increasing the number or the quantity of articles suited to the wants of a whole people, and adapted to promote the enjoyment of that moderate competency which prevails among them.

That observation may have been a touch cold and wordy, but the message was clear: Americans were producing for the common people, not a leisured nobility. That meant mass production and entrepreneurship. In 1852 *The North American Review* observed:

> The Great Exhibition has done more than any thing else,—perhaps, than all else—to illustrate to Europeans the mission of the Anglo-American. . . . Other nations had taxed all their skill to fashion brilliant gewgaws, such as might minister to the pride of nobles and potentates; while we brought an array of machinery, designed to mitigate the toil of the common laborer, and to cheapen the food, the garments, and the furniture of common life.

Americans were proud of their plebeian products, their innovative evolution away from a simple agricultural economy, and they had no problem boasting about it. "Diffuse, then, knowledge everywhere, throughout the length and breadth of this great country," proclaimed *De Bow's Review,* an American magazine devoted to celebrating progress, the year before the Great Exhibition. "[L]et the civilizing and godlike influences of machinery uninteruptedly [*sic*] extend—then will the future of our country, open, boundless and great, beyond all example, beyond all compare, and countless ages bless its mission and acknowledge its glorious dominion."

Some found American individualists to be crude, insufferable braggarts, and dangerous too. The British were enthralled by tales of American ruffians eager to fight with guns or knives. After seeing "the most beautiful cutlery" at the Sheffield display, Queen Victoria wrote in her journal that there "were Bowie knives in profusion, made entirely for Americans, who never move without one."

As high summer arrived, the American contingent at the Great Exhibition started receiving more respect. First there was McCormick's reaper, an agricultural gadget billed as a tool to make harvesting quicker, easier, more thorough, and most importantly, more economical. On July 24, at a rain-drenched farm forty-five miles from London, the McCormick machine proved that nothing the Empire had could match it. The British gave the American reaper a three-cheer salute for its victory. Robbins & Lawrence, a Vermont gunmaking team, was also winning praise for its manufacturing system, with machines cranking out shoulder arms at a pace faster than more traditional makers. For that they could thank state-of-the-art machinery powered by a

Connecticut River tributary that ran past their stone factory, turning a water wheel whose shaft jutted into the building to bring the machines to life. What really got Robbins & Lawrence noticed was that the firm was well on its way to making interchangeable parts.

The exhibit set up by the Vermonters was as plain as the rest of the American offerings. Robbins & Lawrence brought six rifles that could be taken down, their parts scrambled indiscriminately, and put together again without regard to which part went in which gun. This feat astonished the British. Like the Springfield Armory, the Robbins & Lawrence factory in rural Windsor, Vermont, was an incubator of creativity. At the time of the Great Exhibition, Horace Smith was in Windsor, trying to turn Walter Hunt's Volition Repeater into something that would work well enough to be marketable. His future partner, Daniel Baird Wesson, also came to Robbins & Lawrence, where he supervised the manufacture of pepperbox pistols.

While visitors to the Crystal Palace were impressed by Robbins & Lawrence's achievements—the firm would be awarded a medal—the emerging star of the American show in late August was the thirty-seven-year-old Connecticut Yankee with a flair for attention and a patent for fast-firing guns. Ever the self-promoter with a life-long passion for physical things, especially ones that erupted in spectacular fashion, Colt was in heaven at the Crystal Palace.

At his display Colt reveled in showing how he, too, could assemble his guns from random parts. In a nearby cabinet he kept cigars and brandy for the pleasure of favored patrons, as well as himself (Colt enjoyed fine liquor). He also came bearing gifts, such as a pair of engraved revolvers—one an 1851 Navy and the other a third model Dragoon—for Prince Albert. To house these guns Colt had an Austrian firm make a rosewood presentation box lined with blue velvet and inlaid with brass, mother-of-pearl, ivory, and white metal. He gave another pair of revolvers in a similar case to attorney Ned Dickerson as a thank-you for beating the Massachusetts Arms Company in court. Colt would continue to give away many more, either as tokens of gratitude or to curry favor.

By summer's end the Americans had gone from ignominy to triumph. Not only had Colt's and McCormick's innovations demonstrated technological excellence, but Charles Goodyear had shown the merits of his vulcanized rubber, and Gail Borden's meat biscuits had won the Great Council Medal. Borden would later win even more acclaim for condensing milk. A New York locksmith picked a famously unpickable British lock, and the schooner *America* beat all British yachts in a race around the Isle of Wight to win what came to be known as the America's Cup.

In the years immediately following the Great Exhibition, European travelers in the United States noted the growth in the country's population, its rapidly spreading rail and telegraph lines, and the abundant use of machines. "Standing on American ground," wrote one, "I think of the future." Like his country Samuel Colt was also tasting real success on an international level. He would not let it subside. "[I]f I cant be first I wont be second in anything," he once proclaimed, and he was living up to the pledge. Soon he would establish a gun factory in England designed to his specifications and filled with American machinery. The British Empire had once set the standards for industrial production. Now it was the Americans, especially Robbins & Lawrence and Colt's Patent Fire Arms Manufacturing Company, who were beginning to set them. Of those two firms, only Colt's would still be around six years later.

15

PERFECTION IN THE ART OF DESTRUCTION

It was now *Colonel* Colt.

For years, his brother had called him that without justification, just as he had never really been "Dr. Coult" in his laughing-gas years. He had no military experience, his only combat having been in court—if physically thrashing a lawyer who tried to collect a debt doesn't count. But he had backed the successful 1850 gubernatorial campaign of his friend and fellow Democrat Thomas Henry Seymour, a celebrated combat veteran of the Mexican-American War. In apparent gratitude one of Seymour's first acts as governor was to make Colt a lieutenant colonel and aide-de-camp in the Connecticut state militia.

It was an honorary title, one that came with no soldiers to command in battle, but from then on, Colt could legitimately call himself "Colonel." And he expected others to do the same. Colt began attaching the title not only to correspondence but to the address lines on top of some of his revolvers: "ADDRESS COL. SAML COLT NEW-YORK U.S. AMERICA" was one variation. He hired a New York military clothier of the highest reputation to dress him in a wool uniform made from the finest broadcloth, twenty stitches to the inch, topped by an officer's hat sporting an embroidered bugle insignia and a black ostrich feather.

What "Colonel" gave to Samuel Colt was more than a new title to strut behind. It brought him military cachet, a valuable asset when it came to selling guns to armies. He wouldn't need another legendary spokesman like Samuel H. Walker, and wouldn't have to risk his hide in battle, though he would offer to serve, if he could lead troops supplied with his revolving guns. (No one took him up on it.) His success at the Great Exhibition would help even more, and he wasted no time before telling the world about it.

"Among all the objects of interest which the skill and enterprise of our countrymen presented to the world in this great theatre," boasted a company pamphlet the year after the Great Exhibition had closed, "none attracted more attention than 'Colt's Repeating Pistols.'" In the Crystal Palace, the pamphlet continued,

> were found the warriors of all nations—from the Chinese foot-soldier—whose knowledge carries him no farther than to turn summersets to frighten the enemy,—to the splendid trooper of the Horse-Guards, whose clanking accoutrements, glittering in the sun, delighted the more primitive soldier. All these men of war naturally clustered about the stands where implements of destruction were exhibited; the more refined soldier seeing in the greater power of modern weapons the most potent elements of peace; and the more savage only considering how much more revenge he might take upon his half-civilized enemy.

Different though these fighting men were, the company insisted, they all agreed on one thing: Samuel Colt's revolver "had reached perfection in the art of destruction."

So impressed were the British that the Institution of Civil Engineers, one of the most respected professional engineering associations in the world, invited Colt to address its members in November 1851, the month after the Great Exhibition closed. He was the first non-Englishman to speak before the group, which would eventually make him an honorary member.

"[E]xperience has shown," Colt told the engineers, "that perfect weapons of defence are indispensable for the pioneers of civilization in new countries, and still as necessary for the preservation of peace in old countries. . . ."

The best means of producing them, he insisted, was not by hand but by machinery like the industrial set-up in his Hartford factory. Rapid-shooting guns were also valuable for the American pioneer to protect himself and his family "living in a country of most extensive frontier, still inhabited by hordes of aborigines. . . ." The message to Great Britain, a nation with far-flung colonies, was unmistakable: A former colony is benefiting from my product, and an empire shouldn't rest on its laurels.

Attendee Abbott Lawrence, US minister to the United Kingdom and a wealthy textile entrepreneur, was invited to hear Colt's address. General William Harney, who had fought in the Seminole and Mexican-American Wars, liked Colt revolvers, Lawrence said. Speaking for himself, Lawrence said that the American military thought Colt revolvers were "the most efficient weapons ever introduced, particularly for border warfare, against savage tribes, whose cunning, hardihood, courage and skill, rendered them very formidable enemies." At his revolver-making operation in Hartford, thirty thousand weapons were at the moment "in various stages of progress," Colt said. His workforce of three hundred was finishing about a hundred guns every day. Demand was so great, he claimed, that he was thinking about boosting annual production to fifty-five thousand.

Because the revolver business was excellent, especially for guns that worked like a Colt, other makers were coming out with their own versions. Some were far too similar for Samuel Colt's liking, so once again he sicced Ned Dickerson on a couple of men he accused of violating his patent. And again Dickerson triumphed at trial. In case anyone else was foolish enough to think of entering the market with a knockoff revolver, Colt had Dickerson send a warning to people in the firearms trade early in November 1852:

> You will please to take notice to desist forthwith from the sale of any REPEATING FIRE ARMS, in which rotation, or locking and releasing, are produced by combining the breech with the lock; or in which the cones are separated by partitions, or set into recesses; except such as are made by Col. Colt, at Hartford.

> All rotary arms constructed with such combinations, whether made by the Springfield Arms Co., by Young & Leavitt, by Allen & Thurber, by Blunt & Syms, by Marstin & Sprague, by Bolen, or by any other person, are a plain violation of Col. Colt's patent; and I shall proceed against you and hold you responsible for damages, if you persist in the sale of any such arms.

Colt's patent wasn't set to expire until 1857, which gave him plenty of time to arm as many people and militaries who wanted state-of-the-art weaponry as possible. Always in a gift-giving mood when he thought it would benefit him, Colt kept his favors flowing. One recipient of his largesse in 1852 was Franklin Pierce, a New Hampshire politician and Mexican-American War general who had won the Democratic nomination for president. While Pierce was on his way to becoming the nation's fourteenth president, Colt gave him an 1851 Navy revolver in a blue velvet–lined mahogany case containing tools and a powder flask. The gun was elaborately engraved in the "donut scroll" style with thick spiraling leaves; even its screwheads had a touch of flourish, and a wolf's head adorned each side of the hammer. Whether Colt also gave a revolver to Pierce's losing general-election opponent, General Winfield Scott, is unknown, but Scott did receive another kind of promotional gift from Colt: a transcript of the Massachusetts Arms Company patent trial bound in Moroccan leather with gilt lettering.

Colt had made his presence known in other countries even before the Great Exhibition. His agents were now in several European nations, while representatives were scouting out prospects on his behalf in the Americas, including Mexico, the country recently defeated by United States troops, some of whom had carried Colt revolvers in combat. (Mexicans had learned to appreciate the power of a Colt.) He also established a beachhead in England. On New Year's Day 1853 he opened a factory on the banks of the Thames River in London. From there he would send guns to customers, particularly the British military, as well as agents in search of more customers.

Colt also had his eye on Asian markets. His friend Commodore Matthew C. Perry was preparing to take a small fleet of ships to Japan in 1853 to convince that isolationist country—by force if necessary—to open relations with the United States. Free trade was seen as a good thing for the US, a nation whose products could use more markets. And Japan had lots of coal, which America wanted. Protection for shipwrecked American sailors was also a priority, because Japan regarded them as outlaws if they happened to wash up on Japanese soil. Perry was to take gifts for the Japanese, at least partly to show how technologically advanced his country was. Colt was eager to have his revolvers included and told Perry in a letter that he would be happy "to furnish them gratuitously having in view the special object of your mission and the reputation of our government." A hundred Colt revolvers went to Japan with Perry.

Gunboat diplomacy worked. The only shots fired were salutes from the American ships, supposedly celebrating the Fourth of July, though the threat of fighting hung over the expedition in its early stages, exemplified by a white flag Perry gave to the Japanese. A letter accompanying the flag left no doubt about its meaning: If you choose to fight and learn how destructive our cannons are, raise this, and we'll stop shooting. The United States and Japan eventually signed a treaty, and trade began.

Real war, always a friend to the arms business, came Colt's way again in the fall of 1853. A quarrel between Russia and the fading Ottoman Empire was drawing in other imperial powers. Great Britain and France worried that Russia wanted to gain access to the Mediterranean through straits controlled by Turkey. That would threaten their trade routes, so the French and British joined in on Turkey's side in March 1854, and the Crimean War began. The British military at the time lacked enough guns to fight a serious war, which the war in the Baltic was quickly becoming. And their reliance on skilled craftsmen who were in limited supply rather than factories employing average workers meant they couldn't get up to speed as fast as they needed.

The United States was officially neutral, which gave Colt cover to sell guns to both sides, though he told the British he wasn't doing that. His friend Thomas Seymour helped him with the Russians, since he was now US minister to Russia, having been appointed by President Pierce, whom Seymour had backed for the presidency. On the revolver-giving front, Seymour had presented Pierce with a cased Colt Dragoon, and Colt had given Seymour a pocket pistol decorated by master engraver Gustave Young. But nothing could compare with the elaborately engraved, gold-accented guns Colt had made for the opposing parties' monarchs, Sultan Abdulmejid I, who then bestowed on Colt status as a Turkish nobleman, and Tsar Nicholas I. Thanks to Seymour, an admirer of the tsar, Colt and his traveling companion, faithful and useful Ned Dickerson, were given a personal audience with Nicholas, who had been staying at the royal residence outside St. Petersburg but came to the Winter Palace just to meet Colt.

By the time Colt left St. Petersburg toward the end of 1854—he would make more trips later—Russia's navy had ordered three thousand of his revolvers. The following April, Prussia closed its border to arms destined for Russia, which threatened to derail the revolver deal. A way around the problem, Colt decided, was to smuggle the guns in. That might have worked, except the revolvers were discovered hidden in bales of cotton at the Prussian border and confiscated.

Soon after war began, orders for guns—British orders, not Russian—from the scene of fighting in the Baltic arrived at Colt's London factory, where they kept his workforce busy turning out three hundred pistols each week. In a leased three-story brick building, which looked like a barracks in a "sombre and smoky region" of the city, Colt put into action what he had talked about in his address to the engineering society. Machines, not human muscle, performed most of the work, and it was all done under one roof, not by true craftsmen, but by laborers who had no gunmaking experience but were trained for specific, easily learned tasks. "Carpenters, cabinet-makers, ex-policemen, butchers, cabmen, hatters, gas-fitters, porters, or, at least, one representative from each of those trades, are steadily drilling and boring at lathes all day in upper rooms," wrote one journalist after touring the factory. The hard work was done by a steam engine "indefatigably toiling in the hot, suffocating smell of rank oil, down in the little stone chamber below." Colt now had the two most modern arms factories in the world, one in London and the other in Hartford, more advanced and efficient even than the United States Armory in Springfield.

Gunmaking was no longer exclusively a man's job. Colt also employed women, whose daily pay in London was two to three shillings—the same rate children were paid—compared to the men's earnings of three to eight shillings per day. "Neat, delicate-handed, little girls do the work that brawny smiths still do in other gun-shops. Most of them have been sempstresses and dressmakers, unused to factory work, but have been induced to conquer some little prejudice against it, by the attraction of better pay than they could hope to get by needlework."

Samuel Colt was not the only entrepreneur who added women to his workforce in the early 1850s. Oliver Winchester also hired them to manufacture men's shirts in New Haven, as the business with John Davies took off. Mending and making clothes was traditionally more "woman's work" than making guns was, so in Winchester's case the hiring was more natural than Colt bringing female workers into his factory.

For the output to keep pace with orders, Winchester & Davies needed lots of women eager to sew. "SHIRT MAKERS WANTED" announced an 1848 advertisement in the *Hartford Daily Courant*. "100 Good Sewers will be furnished with constant employment in making the Patent Shoulder Shirt, for the New Haven Shirt Manufactory." Three years later the company upped its request: "An Unlimited Number of NEAT SEWERS will find constant employment (and cash payments) in making white and colored SHIRTS, or stitching Bosoms, Collars and Wristbands." The women Winchester & Davies enlisted did their sewing at home, sometimes far from the factory—that is until Nathaniel Wheeler introduced himself to Oliver Winchester in 1852.

A maker of small metal objects, such as buckles and buttons, Wheeler had partnered with the inventor of a machine he thought Winchester could use. Its talent was the ability to sew faster than human hands—much faster.

Put enough of these "sewing machines" to work, Wheeler claimed, and Winchester could boost production while saving money. Appliances that promised the ability to sew had started showing up in recent years, but they had yet to make their mark commercially. Their manufacturers were also battling over patents in court, which made production messy. At first Winchester was skeptical about the machine Wheeler was peddling. "I could not prevail upon him to even furnish cloth for a demonstration," Wheeler was quoted as saying later. "The next day I returned, with some cloth already cut, and my wife demonstrated to him by sewing a shirt together before his very eyes. At this, Winchester was so surprised that he gave me a large order, and within the week had agreed with us to purchase more machines."

One more week, and Winchester & Davies were investors in the Wheeler & Wilson sewing machine company. Then they became the first manufacturer to use sewing machines for commercially made shirts. From a workforce of two hundred in 1847 when Winchester & Davies began, the company had grown to employ four thousand by 1855, an impressive number by itself but even more so considering the increased productivity from sewing machines. Winchester ran the factory in New Haven, while Davies took charge of the warehouse in New York, where sales were made. The factory, now four stories high and covering an acre and a half of ground, was making shirts for the entire country, though mostly for the South and Southwest. Net pay for women in the ironing rooms was three to six dollars per week. Those who worked sewing machines were paid more.

Oliver Winchester had made good bets on both shirtmaking and sewing-machine investing. "When we look at the progress made in Sewing Machines," wrote *Scientific American* in the summer of 1852, "we expect them to create a social revolution, for a good housewife will sew a fine shirt, doing all the seams in fine stitching, by one of Wilson's little machines, in a single hour. . . . We suppose that, in a few years, we shall all be wearing shirts, coats, boots, and shoes—the whole habiliments of the *genus homo*—stitched and completed by the Sewing Machine."

Even abolitionist preacher Henry Ward Beecher, brother of *Uncle Tom's Cabin* author Harriet Beecher Stowe, was agog over the sewing machine when a young woman, about sixteen years old, brought a Wheeler & Wilson to his home. "[W]e firmly believed that some things would never be done by any fingers except human," he wrote, "and eminent among these impossible things, was sewing! Nothing, we were sure, could ever perform that, except the latest and best invention of paradise—Woman!"

After the machine embarked on its mission, its treadle nimbly worked by the young operator, Beecher was thunderstruck. A shirt "streamed through the all-puncturing Wheeler and Wilson about as soon as a good-sized flag, being hoisted, would unroll and flow out to the world," he wrote. "A bundle of linen took its turn, and came forth a collar, a handkerchief, a cap. There goes in a piece of cloth, there comes out a shirt! We were bewildered. Not much was done for some hours in that house but gaze and wonder."

In late June 1852, with his shirt business thriving, Oliver Winchester decided to throw a party. The guests he wanted to honor were the New Haven Shirt Manufactory employees, who gathered in his garden alongside the factory early on a Friday evening. There, under a long grapevine arbor adorned at each end with flags and evergreens, the company's shirt sewers, washers, ironers, clerks, and cutters were invited to sit at a 120-foot-long table decorated with flowers arranged by Jane Winchester and the couple's daughter, Ann Rebecca, now seventeen years old. Those who toiled in service to Winchester were served strawberries, ice cream, and cakes, while the Old Gents Band entertained them with music. New Haven's mayor gave a speech.

When it was Winchester's turn to address the gathering, he told them he had chosen not to begin with the standard "Ladies and gentlemen" but instead called the assembled workers his "brothers and sisters, fellow-laborers in a common cause." Yes, Winchester admitted, he was a faultfinder. But when he found fault, he said, he was using "the great motive power which propels the world in its present rapid course of improvement, as the progressive element of the age." Without that pressure, we would not have fast railroads and steamships or "that most wondrous of human inventions, the Electric Telegraph. . . ." The principle of letting well enough alone, Winchester told his employees, must not stand in the way of progress.

The party's purpose was to celebrate his workforce as well as to inspire them, so Winchester lavished praise on the men and women of the New Haven Shirt Manufactory, presenting silver cups for particular excellence on the job. In wrapping up his speech, Winchester summoned them all to strive even harder in their shared quest to better the quality of shirts:

> Perfection belongs only to the Deity; still, while we are short of that point, there is much room for improvement. Let us, therefore, be united in our efforts and purposes, to devote to our several departments, all the energies we possess, nor be satisfied while a stitch is misplaced, a stain unremoved, or a wrinkle unsmoothed; remembering, that a shirt, however coarse, is an emblem of purity, and as the work of our hands, which are directed by our minds, it is the index to our character, to which the close observer of human nature requires no more certain key.

For the now wealthy Oliver Winchester, the time had come to look beyond shirts for even more moneymaking opportunities. In 1855 another novel machine drew his attention, one that he thought had potential. This time it was a repeating gun using a new kind of ammunition that embodied the future: a cartridge with all the ingredients for shooting in a single package. Horace Smith and Daniel Baird Wesson, the men behind the new gun and the company that made it, were looking for investors. Winchester decided to take a chance.

16

EVOLUTION

The New Englander mechanizes as an old Greek sculptured, as the Venetian painted, or the modern Italian sings; a school has grown up whose dominant quality, curiously intense, widespread, and daring, is mechanical imagination.

THE TIMES OF LONDON, AUGUST 22, 1878

Eureka moments of discovery are not the norm in the realm of mechanical creativity. They do happen, but even breakthrough inventions that receive patents are often the product of multiple minds irrigated by the demands of a particular age and the tools available at the time that enable building on what came before.

The revolving cylinder for a gun, for example, had been in limited use before Samuel Colt improved it. The elder Eli Whitney didn't invent the cotton gin; various forms of it were already in existence; what Whitney did was upgrade the way a machine separated cotton fiber from seeds. Samuel F. B. Morse was not the first person to develop a working telegraph; others, including Europeans and Englishmen, had also been developing it. Morse's contribution was using someone else's method of coiling wires to boost the strength of electric signals.

Certainly Walter Hunt, for all his brilliance, relied on those who came before him, as those who came after relied on him. A full decade before the Patent Office issued Hunt his Volition Repeater patent, three men in northern New England received exclusive rights to a rifle that could feed fifteen cartridges one by one into the barrel from a tube inside its wooden stock. These inventors called their mechanism one of "sliding chambers," each cartridge containing all the essential elements for shooting, including a primer, put together in advance by the shooter. While this repeating gun may have had possibilities—its inventors thought so, anyway—it was not a success.

The safety pin Hunt invented was striking in its mechanical purity. It performed one simple task without complication or multiple parts. But his problematic Volition Repeater was not so pure. It was a complex combination of metal pieces, many of them delicate, of varying shapes and duties that needed to interact in perfect harmony for the thing to work. And that was not happening.

Hunt thought his curious firearm had promise and wanted to see it go into production, but, as usual, he didn't have the resources. Machinist and speculator George A. Arrowsmith did. Arrowsmith took up the challenge of turning the Volition Repeater from a sterling idea into a workable product, and with that, the assembly line of invention cranked to life. Over time, Hunt's creation would pass through the hands of other mechanically inclined men, some of them future gun-industry giants, each with his own résumé of inventing, each adding his own improvements. It was a relay race to perfect a product and then profit from it, a process that needed two mutually dependent sets of actors: men with ideas and men with money they were willing to risk. The race began with George Arrowsmith.

Arrowsmith was well aware of Hunt's passion for inventing and had wanted to reap rewards from it long before the Volition Repeater came along. On December 11, 1845, nearly four years before his multi-shot rifle received its US patent, Hunt assigned to Arrowsmith his patent for a spring-loaded inkwell that opened when the tip of a pen pushed down on it. Once the desired amount of ink was collected, the writing implement would be withdrawn, and the opening would close. Earlier that same year, Hunt assigned patents for two other inkwells to Arrowsmith's son, Augustus. In the mid-1830s, he had also sold George Arrowsmith the rights to his version of the sewing machine, which Arrowsmith failed to patent. Whether he made any money on Hunt's self-closing inkwell is unclear, but Arrowsmith was confident enough about the Volition Repeater's prospects to enlist gunsmith and all-around mechanical wizard Lewis Jennings to work on it.

Like Walter Hunt, Lewis Jennings habitually spun his mind in many inventive directions, and the California Gold Rush likely stimulated it to concoct a gadget for separating gold from worthless gunk. On May 1, 1849, Jennings patented what he called a "Gold-Washer," which forced water containing the precious metal, earth, and other "Foreign Matter" through tubes so that the heavy gold would sink to the bottom. This invention, a kind of reverse mechanized decanter, was an early version of what would later be called an elutriator, which continues to be used today to separate minerals. Then Jennings turned his attention to locks, patenting an improved mechanism for securing bank doors and vaults.

But Jennings's big challenge was the Volition Repeater. If he could work out its kinks, there was hope of real money to be made from the military in the United States and perhaps in other countries eager to take advantage of a multi-shot firearm. Arrowsmith was counting on it. Jennings simplified the mechanism but kept the all-important tubular Rocket Ball magazine beneath the barrel. He worked quickly. On Christmas Day 1849 Jennings received a patent for his redo of the Volition Repeater, which he immediately assigned to his boss, George Arrowsmith. These improvements were enough for Arrowsmith to believe that the moment for production had arrived. Arrowsmith, however, lacked the resources needed for full-scale arms making. His best option was to convince someone up the financial food chain to undertake transforming what was now the Jennings repeater into a mass-produced reality. New Yorker Courtlandt Palmer was that someone.

Palmer was both wealthy and well connected, having made money in hardware. The Panic of 1837 had crippled his business, but he survived and rebuilt his fortune by expanding into real estate. Palmer had also been president of the New York, Providence & Boston Railroad until he was bulldozed out of the way by railroad and shipping tycoon Cornelius Vanderbilt, a man he considered treacherous. Maybe Palmer wasn't as savvy as Vanderbilt when it came to railroads, or as ruthlessly combative, but he had a sense of technology's possibilities. That's why he spent $10,000 (about $350,000 today) to buy from Arrowsmith all the rights to the Jennings repeater, even though he knew little about firearms.

Palmer had the patent for the gun, but he needed a factory to produce it. He settled on Robbins & Lawrence, the relatively new enterprise 250 miles north in Windsor, Vermont, a town far up the Connecticut River with a growing reputation for making machine tools that would be enhanced by its success at the Great Exhibition in London. In its five short years of existence, Robbins & Lawrence had become the largest private small-arms factory in the country—even bigger than Samuel Colt's at the time—surpassed in size only by the two government armories. It had taken a major contract for rifles and turned them out ahead of schedule. The factory's leading figure, inventor and gunsmith Richard S. Lawrence, was known as a cutting-edge innovator who had risen to prominence from nothing. The stories of the Robbins & Lawrence Armory and of Richard Lawrence himself are quintessential tales of nineteenth-century Yankee capitalism. Both would play roles in the development of repeating firearms. And, like Samuel Colt, their salvation would be war.

Richard Lawrence was nine years old when his father died, and it fell to him to help support the family. At first it was farm labor on family acreage near Watertown, New York, with no time for schooling, except for learning to make carpenters' and joiners' tools at a woodworking shop. Lawrence was moderately successful in his duties, handling tools with ease and efficiency, but what really intrigued him were the doings in the shop's basement. Custom guns were made there, and that's where Lawrence spent his spare time. He soon developed expertise in repairing firearms.

In 1838, after a brief and uneventful stint in the Army, twenty-one-year-old Lawrence arrived in Windsor looking for a new start. A local doctor named Dyer Story owned a rifle that needed work and asked Lawrence, who professed to know something about guns, if he would repair it. Lawrence took the rifle apart and performed an assortment of tasks with a dexterity that astonished those who watched him. He installed a peep sight—a disk with a hole in the center—on top of the barrel close to the shooter's eye. A typical rear sight tended to be a V-shaped groove. The shooter would look down the barrel and nestle the distant front sight into the groove. Lawrence's circular sight was something no one in the area had seen before.

When work on Story's rifle was done, the two men went out to test it. Their target was a hole drilled by an auger in a maple tree, formerly the site of a sap spout. It was agreed that Lawrence would do the shooting at a distance of about two hundred feet.

"The Doct. tended target," Lawrence recalled. "Could find no ball hole. Said I had missed the tree. I fired again—no ball hole to be found. Doct. came up to me and said I had spoiled his Rifle." Lawrence apologized profusely, promising to make the gun work properly if given another chance.

> He said he could not consent to my doing anything more to improve the shooting qualities—the sight he liked very much. I said that as the gun was loaded [I] would take one more shot and see if I could not hit the tree. After the third shot I went up to the tree to investigate, and all of the three balls which I had fired were found in the auger hole.

The doctor was astounded. Never had he heard of such shooting. The two men spent half that night talking about guns. The next day, Story took Lawrence to visit N. Kendall & Co., a firm that was making guns at a local prison, using inmate labor for the rough tasks and free men for the more intricate work. On the doctor's recommendation, Lawrence was hired immediately and set about fitting stocks on barrels and actions by hand. At the end of six months Lawrence had mastered everything about the little factory—every task, every assignment, every challenge—even the intricate art of engraving. Because of his success, he was put in charge of the shop.

After a couple of years away from gunmaking, Nicanor Kendall and Lawrence opened a small custom gunmaking operation in Windsor. Business was good, but by the winter of 1844, it would get the chance to be even better. Their soon-to-be associate, Samuel E. Robbins, a Boston businessman who had made a fortune in lumber and retired to the Windsor area at the age of thirty-three, told Kendall and Lawrence he had learned that the government was in the market for ten thousand new rifles. That kind of order was well beyond the capacity of the little business, so Robbins—who wanted to use some of his capital in a new venture—had a proposal: Set up a partnership, with Robbins as one of the partners, and send a bid to Washington. They'd see what would happen and make their next move accordingly.

The government accepted the offer: ten thousand rifles at $10.90 each to be finished within three years. Now it was up to Kendall, Robbins & Lawrence to deliver. Unfortunately, the partnership had no machinery that could handle such a huge job. Like Samuel Colt when he got his contract for Walker revolvers, the partners didn't even have a large enough building or a place to put one. So they bought land along a tributary of the Connecticut River, whose strong flow could turn the massive wheels needed to work large factory machinery. Then they built their factory—a three-and-a-half-story brick building a hundred feet long and forty-five feet wide, topped by a weathervane in the form of the rifle they had contracted to build—and either made or bought machinery they would adapt to the tasks under Lawrence's direction. Their Windsor factory drew craftsmen from all over the region: machinists from the Eli Whitney Armory in New Haven, machine shops in Massachusetts and Connecticut, the Springfield Armory, and even as far away as the Harpers Ferry Armory in Virginia.

Lawrence, whose visionary but practical genius guided everything, ensured that this enterprise, larger than anything in Windsor, performed with the precision of a fine watch. Gun stocks were not carved by hand but by a machine whose belts, gears, shafts, and cutters made identical wooden copies of an iron original much like the system invented by Thomas Blanchard at the Springfield Armory. Gauges made sure that metal parts were the same from one gun to the next. Not only were rifles made and made well there but all ten thousand came off the assembly line eighteen months ahead of schedule and at a fine profit. The United States government was pleased enough to give Robbins & Lawrence—by then they had bought out Kendall—a second contract, this time for fifteen thousand rifles.

Because the Vermont firm had the talent, space, and machinery for the kind of output Courtlandt Palmer wanted for his rifle, not to mention the armory's admirable reputation, he contracted for Robbins & Lawrence to build five thousand Jennings repeaters mere weeks after receiving the patent rights. Lewis Jennings had done his best to make Hunt's rifle worth shooting, but it wasn't good enough. The thing was so buggy that Robbins & Lawrence made only about a thousand, including repeaters and single-shot versions. There were also some muzzle-loading Jennings rifles—in effect, a step backward in evolution.

So far Palmer's investment in the future of firearms was a loser. He had sunk a fortune into what he thought was a reasonable risk, only to see nothing come of it other than some fancy machine work and a few underpowered, complex guns no one wanted to buy. But with so much at stake, Palmer wasn't ready to give up. If he put just a bit more money toward making the rifle perform properly, the rewards could be enormous. What he needed was another gifted machinist who could pick up where others left off, someone with a fresh perspective who dreamed boldly but remained reasonable and focused. His choice was a good one. Palmer's new idea man was Horace Smith.

When he left Allen & Thurber's factory in Norwich to go out on his own, Horace Smith continued doing what he was best at: making firearms. He stayed in Norwich working mostly on shotguns, though he found a market for something new: whaling guns.

It was whaling's golden age. Along the New England coast towns sent ships sailing throughout the world's oceans to harvest precious oils from within the leviathans' bodies. Whale oil lubricated the growing number of machines on which the Industrial Revolution depended, burned brightly in lamps, lighthouses, and locomotive headlights, and became soap. Spermaceti, a white wax-like material taken from sperm whales' heads, made the best candles, and ambergris from the animals' intestines stabilized the scents of the finest perfumes. Whalebone became corsets, collar stays, toys, and buggy whips. Hunting whales was big business in the middle of the nineteenth century, especially for the United States, whose merchants built the largest fleet of whaling ships in history. Only textiles and arms making topped whaling as New England industries in the 1840s. Whaling was also a pioneer in high-risk, high-reward venture capitalism, with agents connecting wealthy investors interested in buying ships and hiring crews in a system where everyone had a stake in potential profit. But whaling was also dangerous business for harpooners, as well as the captains and mates who wielded the killing lances. They had to come alongside the giant mammals on the ocean surface in narrow thirty-foot boats and thrust their weapons as hard as they could into the whales' sides. Far better from the whalers' point of view would be hurling their missiles from a distance, which was where guns came in.

A dozen miles south of Norwich was New London, then one of the three busiest whaling ports in the world, along with New Bedford and Nantucket in Massachusetts, so it was natural for Horace Smith to use his talent to make a harpoon gun that would find a place in the whaling industry. A fellow Norwich gunsmith, Oliver Allen, who may have worked with Smith, decided to go one step further and make a lance that would explode inside the whale. Such a "bomb lance" needed a powerful gun to launch it, so Allen made them heavy—twenty pounds or more—each with a bore large enough to hold a shaft that could be plunged deep into a whale's body. Gone was some of the romance of thrusting iron from mere feet away, but it was safer for the men doing the killing. Supposedly Horace Smith also invented an explosive bullet for whaling. If so, there seems to have been no patent for it.

Allen didn't stay long in Norwich. News that gold was easy pickings in California started a countrywide epidemic of gold fever that didn't spare East Coast whaling towns. Men who would have sought their fortunes at sea or in businesses that benefited from whaling headed instead to the gold fields of California. Among them was Oliver Allen, who joined sixty other gold seekers in 1849 to buy a ship and sail around Cape Horn on their way to the Pacific coast. Allen would spend the rest of his life in California.

Horace Smith also moved on, but not with a journey to far-off gold fields. Whaling guns weren't working out for him, but he kept making other firearms in Norwich, now assisted by his teenage son, Dexter, who had at least two of his father's qualities: a talent for mechanics and a quiet disposition. The elder Smith also made repeating guns, though not of his own design. A pair of inventors who had patented an oddball firearm with separate compartments for powder and balls paid Smith to make a few of them, which he did. Those guns went nowhere in the market.

Then came venture capitalist Courtlandt Palmer's enticement to tackle the Jennings gun for a salary and a share of the profits. Smith simplified what Jennings had done and resurrected the tubular ammunition magazine, which was the main component that allowed the gun to be a repeater. Less than two years after Lewis Jennings got his patent and production on his model began, Smith had his own patent for what would be known as the Smith-Jennings rifle. But the gun still had its problems. One was something that had bedeviled it from the beginning: The Rocket Ball, while a harbinger of things to come, was pathetically puny. It lacked the clout one expects from a serious gun. Also, a separate explosive pill that had to fall into place for each shot slowed things down, its residue occasionally fouling the mechanism. And the gun's complicated and unreliable mechanism, though better than when it came into Smith's hands, contributed to its demise.

By 1852 it was time to give up. The mechanical descendants of Walter Hunt's Volition Repeater had almost been worth making. For sure they were closer to finding a place in the market than their ancestor had been, but not close enough, despite Courtlandt Palmer's money and the creative minds of Lewis Jennings and Horace Smith. Even Richard Lawrence and the talented Robbins & Lawrence shop foreman, Benjamin Tyler Henry, usually referred to as B. Tyler Henry, had made suggestions for improvements from the beginning, but these didn't help much. Success seemed just out of reach.

For Palmer it appeared to be the end of the line. His eagerness to spend money in the hope of great rewards was gone. Just in case, he did hold on to the patent that Horace Smith had assigned to him. As for Smith, he would continue to tinker with the troublesome mechanism that he knew had possibilities. Henry would stay on at Robbins & Lawrence until lured away by another venture capitalist who, like Courtlandt Palmer, knew little about guns but was ready to put his cash on repeating firearms in the lineage started by Walter Hunt. This money man was Oliver Fisher Winchester.

Over the years creative talent poured into Robbins & Lawrence, and new manufacturing methods flowed out. Christian Sharps, whose breech-loading carbines and rifles would be carried by Union troops in the Civil War, worked with them. So did J. D. Alvord, who applied his Robbins & Lawrence learning to help Wheeler & Wilson make sewing machines. Workers trained at the backwoods armory in northern New England fostered innovation in firearms, metalworking, machine tools, shoe machinery, and railroads. But despite the company's success and the plaudits it received at the Crystal Palace, problems with a British contract for guns to be used in the Crimean War, plus an unwise venture in railway cars, crippled it. Robbins & Lawrence went bankrupt in 1856, a decade after it was born.

Still, the gun of Walter Hunt's imagination continued to evolve in the hands of Horace Smith, who would combine forces with another inventor interested in repeating firearms, a younger man named Daniel Baird Wesson, who would become his business partner for life. With the help of Wesson, whose mechanical creativity Smith had come to respect, perhaps making a repeating rifle would finally turn a profit.

17

A NEW ORGAN OF DESTRUCTIVENESS

Exactly when and where Smith first met Wesson is debatable. It might have been at Robbins & Lawrence when Smith was trying to make something out of the Jennings repeater. They certainly had the opportunity. Wesson was there at the time supervising the production of pepperbox pistols for his then-employer, the Leonard Pistol Works, which had contracted with Robbins & Lawrence to make their guns. Or they may have met when they were Massachusetts Arms Company stockholders during its ill-fated attempt to make Wesson & Leavitt revolvers.

Most likely the two became acquainted with each other at Allen, Brown & Luther, a gun and toolmaker in Daniel Wesson's hometown of Worcester, Massachusetts. Both men needed work, and thriving Worcester was a logical destination, even though for Smith it was more than fifty miles from Norwich. With inquisitive minds, each craving to invent a workable repeating firearm that would sell, Horace Smith and Daniel Wesson were natural colleagues. Yet their histories and temperaments were very different.

Smith was more than sixteen years older than Wesson and had suffered twin tragedies of losing both his wife and infant son when Wesson was still a child. As a teenager he had learned over long years how to tame metal within the vast order of the government armory at Springfield, where machines dominated everything and everyone, to make weapons for the military. Smith's time at Eli Whitney's armory, another big enterprise feeding off government work, also schooled him in large-scale production. Wesson, on the other hand, learned gunmaking from an elder brother in a private shop with a clientele of discriminating individuals. Perhaps his history of steady, often tedious work made Smith the serious man he was, a focused individual whose approach to problems was methodical and deliberate. Wesson was more dynamic, with the enthusiasm of youth as well as a touch of his brother's ebullience. Thus it made sense for Smith to be the senior partner and not just because of the age difference. Understanding that their strengths complemented each other, Horace Smith and Daniel Wesson formed a loose partnership in 1852 in Norwich, the Connecticut town where gunsmithing thrived and Smith had first tried to make it on his own. Together they applied their talents to the single goal of turning the latest descendant of Walter Hunt's Volition Repeater into something that would shoot quickly, often, reliably, and with power. They wanted to succeed where others had fallen short.

The solution, they decided, was less in the gun than it was in the cartridge. First the lackluster Rocket Ball did not live up to its name. Next the business of putting a small explosive pill behind the bullet that housed gunpowder was not always smooth, despite whatever ingenious device ladled it out, and sometimes it failed. Getting the pill into place was also an extra step that slowed down the process of readying the gun for the next shot. Combining everything—projectile, gunpowder, and primer—into a single compact package would be a great improvement. Even better would be a bullet that landed a bigger punch than did the anemic Rocket Ball.

As usual in the chain of human ingenuity, neither Smith nor Wesson was the first to come up with an all-in-one cartridge. It was the Europeans who led the way. Jean Samuel Pauly, a Swiss living in Paris, started it all early in the nineteenth century with a cartridge made of paper or metal on a copper base. Then in the 1840s a Parisian gunsmith named Louis Nicolas Auguste Flobert designed a cartridge containing volatile fulminate inside around the rim of the base. Flobert's invention lacked power, but that didn't matter to his customers, since they used it only for the mild sport of indoor target-shooting with single-shot "parlor" pistols, also made by Flobert. They were, in effect, a kind of toy. The Flobert cartridge contained no gunpowder for the fulminate to ignite, so the only propellant was the primer. That meant the bullet had hardly any power behind it. This wasn't a problem for shooting at targets indoors; in fact, the Flobert's gentleness was an asset for those wanting minimal noise and recoil. But for the military, killing power was essential. The Flobert wouldn't do for combat.

Wesson didn't care much about the Flobert pistol, though he ended up making a few. What interested him was its ammunition. Put all the components required to shoot into one metal package, line up several of those packages in a mechanism that fed them one at a time into a chamber at the back end of a gun's barrel, and you had a repeating firearm. That package, called a cartridge, was the key to success. Europeans had been experimenting with various cartridges for a while with some progress. One type that was catching on was the so-called "pinfire." As the name implies, it fired when the gun's hammer struck a pin protruding from a copper cartridge. The pinfire never fared well in the United States, partly because the exposed pin made accidental discharge a hazard, especially if someone carried several together in a pocket.

Still, the Parisian's little cartridge held promise for Smith and Wesson. If it could be beefed up—it was too delicate to handle a serious powder charge—and kept to an all-in-one package, maybe they could modify the Smith-Jennings rifle to shoot it. Smith and Wesson saw so much promise in the Flobert's future and were so intrigued by the possibilities that they devoted virtually all their time to coming up with the perfect combination of cartridge and repeating rifle. Making guns to sell took a back seat to experimentation. Courtlandt Palmer had dug deep into his pockets to gamble on a new type of rifle. Now it was Wesson and Smith who were taking a financial risk.

For most of 1852 and into 1853, the partners worked on correcting problems with the Jennings rifle. Wesson tried a heftier cartridge based on the Flobert, but when it wouldn't work well in the gun, he returned to the Rocket Ball, this time with a primer in the base rather than separate. He and Smith filed for patents on cartridge improvements and on a new gun they had fashioned that fed loads into the chamber and cocked its hammer all in one motion by using a lever that doubled as a trigger guard. It would take months before the government decided whether to issue the patents they sought. In the meantime, with little money coming in, the partners were feeling financially squeezed.

Then a rescuer arrived—or rather, returned. Courtlandt Palmer was not done with the problematic repeater after all. Cornelius Vanderbilt may have been right to think that he caved under pressure, but Palmer was tenacious when it came to the repeating rifle that he had propped up with his capital. He still had hopes for it, as well as faith in the talents of Smith and Wesson, so he struck a deal with the partners. In exchange for licensing the rights to the Hunt, Smith, and Jennings patents, Palmer would be a formal partner with Wesson and Smith and take most of the profits sure to flow from the gun the inventors finally came up with. Actually, had Smith and Wesson not made a deal with Palmer, they might have been on dangerous ground if whatever they invented came close to the patents owned by the New York capitalist. At least now they were safe. Palmer also threw in $10,000 cash to get the new enterprise going. He would stay in the background, leaving center stage to the two inventors working together in Smith's Norwich shop.

In August 1854 the Patent Office issued the gun and cartridge patents to Smith and Wesson. Now the challenge was to convince the public that they had a weapon worth buying. One thing going for the gun was its nickname: the Volcanic, which brought to mind a force of unrelenting power, an eruption of bullets. It was perfect for publicity, even though its Rocket Ball projectiles were less than potent.

To push its product Smith and Wesson reached out to the press and found a friend in the *New Haven Palladium*. In early 1855 the paper gave them the publicity they wanted. "We have seen and fired a pistol, recently invented and patented, which bids fair to excel everything in that line that has yet been offered to the public attention," The *Palladium* raved under the heading "Volcanic Repeating Pistol." This gun, the paper continued, "seems to combine all that could be desired in such a weapon." Then it took aim at the biggest name in guns, the man who had killed Daniel Wesson's attempt to bring his late brother's revolver into the marketplace: "Colt's pistol compared with it seems like a distortion, or a clumsy, uncouth, and ridiculous affair for a fire-arm."

What the *New Haven Palladium*—or rather the Smith and Wesson partnership—had to say about the supposed marvel of the Volcanic was picked up word for word in newspapers from Bangor, Maine, to Shreveport and Baton Rouge, Louisiana; to Platteville, Wisconsin; Detroit, Michigan; and Louisville, Kentucky. When a shipment of Volcanics arrived by steamship at Charleston, South Carolina, in February 1855, a local hardware merchant advertised that they could "be discharged with greater rapidity and certainty, than any other Pistol, now in use."

At the end of March 1855, a Smith & Wesson agent was in Wisconsin touting the Volcanic's excellence. "This is a new 'organ of destructiveness,' which rather exceeds in rapidity and efficiency of execution any thing we have yet seen," declared a promotion in *The Milwaukee Daily Sentinel*. Referring to the new self-contained cartridge, the notice said, "The 'ball, powder and priming' are all done up together in a neat-looking pill—though like other pills, mighty unpleasant to take. . . ." A couple of months later, a prominent hardware dealer told Milwaukeeans that he, too, had these marvelous new guns to sell. "THE SUBSCRIBERS HAVE JUST received a small lot of the justly celebrated Volcanic Repeating Pistols," announced H. J. Nazro & Co. several times in late May and early June 1855. "*Thirty charges can be loaded and discharged in 50 seconds*. The ammunition is Water-Proof, superior to anything yet out." Despite all the promotional bluster, Volcanics failed to jump off store shelves. Sales were slow. Bugs still needed to be fixed, and the weak Rocket Ball wasn't the hard-hitting bullet customers wanted.

The Volition Repeater's patent was less than six years old, yet it had already undergone many changes. Some of the changes were quite clever, but all ended up as unsuccessful attempts to turn the troublesome repeater into a weapon worth producing. Further efforts to salvage the latest descendant of Walter Hunt's creation required money, and that was running low. Then in early July, a collection of investors, mostly from New Haven but with a few from New York City, stepped up with an offer to take over the patents and production. These weren't gun people but capitalists eager to put their money into ventures that promised a profit. So far, the Volcanic didn't seem to have much promise. That didn't deter the clockmakers, shipping merchant, saddle manufacturer, carriage maker, railroad conductor, and others ready to gamble on the gun. (Chauncey Jerome thought of putting some of his clock money into the venture but decided not to, choosing to merge his company with one owned by showman P. T. Barnum. That business then failed, ruining Jerome financially.) Also hot for this gun investment was a New Haven businessman who had earned a fortune making shirts and wanted to branch out. He knew little about firearms, but he saw potential for making money in them. Before long the mere mention of his name would summon the image of a lever-action rifle.

When the summer of 1855 began, Oliver Winchester was comfortably set. His shirt factory was doing splendidly, especially with sewing machines boosting production in New Haven and in the homes of women with whom he and John Davies had contracted. Forty-four years old and the driving force behind the company that had made him rich, Oliver Winchester was a man of status. Two years earlier he had joined a committee to lobby against using the railroads to deliver mail between New York and Boston on Sundays, "a wholly unnecessary violation of that day of rest, which the great body of American people honor with religious observance." Winchester, the thinking went, was among those with enough clout to seek "redress of this grievance, so that it may be determined whether our Sabbaths are henceforth at the mercy of a few great moneyed corporations," meaning the railroads. (Sunday mail was on its way out then anyway.) He had also been elected to the board of the New Haven Water Company, whose job was to come up with ideas for the industrializing city's first public water system.

What spurred Winchester to heave a chunk of his fortune into guns—especially ones with a poor economic track record—is a mystery. The failure of Smith & Wesson and its predecessors to make money despite a prominent capitalist's backing was not something easily overlooked by a seasoned, well-connected businessman like Winchester. Was it an insatiable urge to embark on a radically different venture? Whatever the reason, it was more than idle dabbling. Not only did Winchester invest cash in the Volcanic Repeating Arms Company at its creation in July 1855, but he also became the firm's vice president.

As part of the deal to take over everything to do with Volcanics, the new company allocated some stock to Smith, Wesson, and Palmer and also agreed to pay them cash installments into 1856. What Palmer received did not come close to what he had invested in the gun, but it was at least something. The Volcanic company acquired all the patents, including the one for the magazine gun issued to Smith and Wesson only the year before. Then the investors kept spending, particularly for new machinery, but one problem remained. Their products were the same gun and ammunition that had fared so poorly in the marketplace when Horace Smith and Daniel Wesson controlled the business. After a few months, the company decided it was time to leave Norwich and set up shop nearly fifty miles away in New Haven, where most of the investors lived and worked. For Smith that was a sign to leave the company and remain in Norwich. Wesson decided to stay on as plant superintendent and move to New Haven with the company.

The Volcanic firm wasn't doing any better than its predecessors at selling. In late 1855 it sent an agent, Joseph W. Storrs, to St. Louis, Missouri, a fast-growing center of trade that Storrs called "the best Pistol market in the country." A half-dozen gun stores there did enough business to be called wholesalers, selling "upwards of Sixty Thousand Dollars worth of Colts Pistols per year." Yet even in gun-hungry St. Louis, Storrs couldn't convince commercial sellers to take a chance on the Volcanic. The problem, he told company secretary Samuel Talcott, was that the big dealers felt entitled to buy at a discount, whereas his directive was "to sell to all at the same price." If this company were his, Storrs said, he'd cut the dealers a break in order to get their business, "as there are thousands of customers they can reach that will not be reached in any other way." Storrs did see some Volcanics that had been brought in for repairs, but the shops "make bad jobs of it," which hurt the gun's reputation. If the major shops sold the repeaters, Storrs told Talcott, "they would keep the material for repairs and do it to the satisfaction of all. They would use their influence to make a reputation for them while now it goes the other way." Company policy was not his to make, Storrs recognized, but as a warning to Talcott, he wrote, "I feel safe in making the prediction that unless a diferent [*sic*] course of policy is persued [*sic*] in two years the thing will prove an entire failure."

The company knew it needed those sales. Investors had spent a lot of money setting up the business, which now lacked sufficient operating reserves to keep it going unless the guns sold well. If the company wanted to stay alive, it had to borrow from shareholders, especially Vice-President Oliver Winchester and its president, shipping merchant Nelson H. Gaston, who had made money in mining. These loans were secured by mortgages on the company's assets, which meant that Gaston's and Winchester's ownership stakes would increase if the firm went under.

Things got worse. In February 1856 Daniel Wesson resigned, leaving no one in charge who was truly knowledgeable about guns. The company began borrowing not only to cover payroll but also to retire debt, all while sales remained sluggish. Then something happened that opened the door for Winchester to start taking over the business, not despite its financial woes but because of them. On December 23, 1856, Nelson Gaston died unexpectedly at the age of fifty-two. Winchester became president and then made a bargain with the executors of Gaston's estate. They would force the Volcanic Repeating Arms Company into bankruptcy to protect their investments in case outside creditors filed claims against the business. That opportunity came on January 10 and 13, 1857, when notes the company owed Winchester and Gaston came due. Without another loan, the company could not pay on the notes, so Winchester and Gaston's executors went to court to foreclose on their mortgages, a move that forced Volcanic into insolvency a month later.

If there was any thought that Oliver Winchester was a mere dilettante and not a shrewd businessman who could play the corporate game, his next actions dispelled it. He lent enough money to the company's bankruptcy trustees, who now controlled Volcanic's assets, to pay off the firm's other loans as well as its remaining installment obligations to Smith, Wesson, and Palmer. Wiping out the debts to Palmer and the two inventors meant that the patents had been completely paid for and were now under the trustees' control. Winchester then bought the right to redeem Gaston's mortgage and acquire various other assets—including two thousand guns in the process of being finished, for a dollar apiece. In the end the sole owner of what had belonged to the dying Volcanic Repeating Arms Company, including its precious patents, was Oliver Winchester. His next step was to form another new company, one that would do exactly as he pleased.

It didn't take long. On May 1, 1857, less than three months after the Volcanic company went under, the New Haven Arms Company rose from its ashes. At the helm stood Oliver Winchester—president, treasurer, and largest shareholder. Over time and with more skillful maneuvering, he would exert all but total control of the company. By now Winchester knew a lot more about guns than he did when he first poured money into the Volcanic Repeating Arms Company. He saw clearly what had all along been wrong with the generations of guns descended from the Volition Repeater. It was the ammunition. Horace Smith and Daniel Wesson had also seen the problem but hadn't managed to solve it, though Wesson came close before returning to his improved Rocket Ball. Winchester knew more thinking had to be done about both the gun and what it shot. He realized that he wasn't the one to do the mechanical brain work, but he knew someone who could.

B. Tyler Henry, the gifted Robbins & Lawrence superintendent who had a hand in developing what became the Smith-Jennings rifle, was just the man for the job. The day Winchester took over the presidency of the New Haven Arms Company was the day Henry came on board. Winchester was banking that Henry's changes to the gun and its ammunition would result in something buyers welcomed, particularly the military with its deep pockets.

Smith and Wesson had not been idle since leaving the Volcanic Repeating Arms Company. Wesson had returned to working on a version of the Flobert cartridge to be shot in a revolver once the Colt patent was no longer an obstacle. He had also built a wooden model of a revolver of his own design that he thought would suit the cartridge well. What he and Smith needed was a way to break into the market with a gun no one could challenge. A rock solid patent would accomplish that. They would soon have one, not of their own creation but from the mind of a man named Rollin White, who once made revolver parts in Samuel Colt's factory but was now selling clothes in Hartford. Their timing would be perfect.

18

IN KING COLT'S SHADOW

Rollin White was only about ten when he first saw a gun that didn't load from the muzzle. The pistol belonged to his father, Josiah, and it had a flintlock mechanism, a system on its way to obsolescence in the late 1820s. The boy was attracted to a curious feature of the little gun: Its barrel unscrewed at the breech, so that it could be loaded from the rear. Of course, the barrel had to be screwed back on before the gun could shoot. An interesting idea, but not a step toward rapid fire. Still, Rollin White would remember this feature as he grew older, eventually applying it in his career as a gunmaker, securing a patent, and enticing the hot pursuit of Daniel Wesson.

It's not clear why Rollin White's father possessed this unusual, though far from unique, pistol. If Josiah had an interest in guns, it wasn't professional, for he was a farmer who owned one of several sawmills, a gristmill, and a shingle mill in Williamstown, a settlement of about thirty-five homes in the hills and valleys of Vermont's interior. Despite soil that was mostly clay with some parts loam or sandy, Williamstown was a prosperous community, and White possessed both property and influence. He owned a pew at a local meetinghouse, and the furnishings in his home outclassed the milieu of rural New England.

Winters were hell on Josiah's sawmill. The building could withstand cold and snow, but like other New England mills of the 1830s, its machinery depended for power on waterwheels turned by the flow of a stream.

Hard Vermont winters slowed all but the strongest streams, and the one that powered Josiah White's mill didn't have that kind of power. Losing profits when the flow dropped to a trickle was not to White's taste, so he decided to harness nature.

His solution was straightforward. The stream his mill used was just below eight-acre Lime Pond, whose stored waterpower could be an effective reservoir of force. Josiah figured that by cutting a break in the barrier that kept the pond's contents in place, he could liberate the water when he needed a boost of power for his mill's wheels. But when he opened a twelve-foot trench, more water rushed out than he had expected. The deluge tore up trees along the stream's banks and sent boulders tumbling toward the mill below. Luckily White's man-made flood spared his mill.

Josiah White may not have had a stake in professional gunmaking, especially given his varied other ventures, but his eldest son did. Josiah Dennis White, known as "J.D." took up the craft as a young man in Williamstown, making pistols, rifles, and shotguns. After brief employment at a brother-in-law's saddle shop and then a sawmill, where he was hurt in an accident, and a gristmill owned by his father, Rollin began working for his brother. As soon as he saw a multi-shot pepperbox pistol in J.D.'s shop, Rollin thought of making a repeating handgun.

"I suggested to my brother," Rollin recalled years later, "to cut off the barrel in front of the breech, for the purpose of loading it in the rear end of the barrel."

That was apparently the end of his thoughts on repeating firearms for several years. Rollin White made no model of such a gun and doesn't seem to have put any design on paper. Instead White became a store clerk, first in Williamstown and later in Boston, before returning to his hometown when his father fell ill. Back in Williamstown, White began serious inventing but not with guns at first. In 1842 he patented a loom for weaving bolting cloths and went to Barre, Massachusetts, where he put two of them in operation. After less than a year in Barre, the peripatetic White again returned to Williamstown, apparently in fine financial shape, because he was able to buy both a gristmill and a sawmill.

Unmarried but available, well established in the town where he was born, Rollin White could have settled into a comfortable life. But by February 1849 he was back in the gun business, this time at Samuel Colt's Hartford factory nearly two hundred miles due south of Williamstown. And he'd brought with him a still-simmering idea for a repeating pistol that could be loaded from the rear.

Rollin was not the first White brother to come to Hartford seeking his fortune. J.D. was there to greet him, as was Mason White, who was seven years younger than Rollin and in the gun business with J.D. But they didn't make their own guns; they helped Samuel Colt make *his* guns. Rollin White wasn't hired for his nascent inventorship. He was just a worker whose job was to "turn" revolver barrels for his brothers in service to Colt's growing gun empire on the banks of the Connecticut River. Turning meant putting each barrel in a lathe and making sure that the cutting process resulted in something that was uniform and up to Colt's standards. Doing that right required a measure of skill, but it was repetitive, not creative. Rollin stopped working for Colt for a couple of months in 1849 but was back at it in August, turning and then polishing barrels while his brothers did the rifling, a task that required a bit more skill than turning did, but, like many factory jobs, it was still repetitive.

There was plenty of work to keep the White brothers busy. Orders for Colt revolvers came pouring in. The California Gold Rush drew adventurers and fortune hunters westward, either by ship around the tip of South America or by wagon across the plains and mountains, with a gun considered essential to all but the foolhardy. A revolver bearing the name Colt was the perfect companion, not only for protection against the dangers of the journey but for defense when in the lawless hellholes Easterners expected to find when they reached their destinations. Having a Colt within reach might even give a hesitant man enough courage to head west. Business was so good that in 1849 Colt moved his operation from the small factory when he started making his post-Walker guns to larger quarters in Hartford. Seventy men toiled for Colt there. Many more would soon be added to the workforce, including a mechanical genius who would make his operation more efficient, more productive, more modern than almost any other large enterprise. That genius was none other than Elisha K. Root, who had rescued teenaged Colt from the angry, mud-soaked July 4th crowd in Ware, Massachusetts, way back in 1829. Root contributed not only a sense of big business organization but also a number of inventions that accelerated the evolution of mass-production and metalworking technology.

Rollin White made a living working for Colt, but he wasn't making satisfaction for himself. He was in the right place to hone his metalworking skills, so he spent his free hours innovating on his own, trying to make a revolver that loaded from the rear, something he hoped would be an improvement on Colt's guns. Loading it that way, he thought, would make the process smoother, and the gun would be easier to keep clean. White was also in the perfect place to find raw materials. He could have bought revolvers to experiment with, but why bother, when there were so many parts lying around his workplace? He chose a couple of Colt cylinders, cut them in half on his company lathe, using a tool he had borrowed from a fellow Colt employee, and put together the front ends of each to make a single cylinder that was open all the way through. He drilled out the rear of each chamber, where a lead ball could be inserted and squeezed to a halt at the front end. This would keep the ball in place until it was blown out the barrel by ignited gunpowder. White called the pair of cylinders he started with "refuse" or trash, because they weren't up to the standards the company required in guns leaving the factory. Both would have been recycled or considered scrap metal, so they weren't really White's to take. Never mind, they wouldn't be missed.

White's next step was to see if his new cylinder worked. For that he needed the rest of the gun, so he took apart a borrowed Colt and reassembled it with his custom cylinder. Now, for somewhere to shoot it without causing too much attention to himself, because, he claimed later, he feared that his employer might not like him experimenting with new ideas for revolvers that could be competition. Ferdinand Steele, a fellow lodger at his boardinghouse, told White he knew just the spot in the Phoenix Iron Works, where he was employed, a business owned, as it happened, by George S. Lincoln, son of an accomplished inventor and one of the men who had lent money to Edwin Wesson. And it was nearly opposite their boardinghouse. In the foundry's basement under the main building, Steele said, some loosely laid floorboards could be moved, allowing White a clear and undisturbed shot into a Connecticut River tributary flowing directly below.

Steele offered to escort White down into the foundry basement after working hours to see whether the modified revolver would perform. At his boardinghouse White packed a cylinder chamber with a lead ball and gunpowder. A major challenge was to keep any flames at the rear of the chamber from calamitously setting off other loaded chambers. For his first experiment White loaded only a single chamber. If gunpowder residue found its way beyond that one, it would show that the others might have gone off. No sense in courting an unintentional fireworks display with bullets going every which way. It was a wise decision.

White put the rest of the gun together before walking to the Phoenix Iron Works for the test. After moving basement floorboards to expose the river below, White cocked the pistol, put the loaded chamber in line with the barrel as best he could with his thumb and finger, pointed the muzzle down toward the river, and pulled the trigger. The gun did his bidding, thunderously sending a bullet into the water. But so much explosive gas blew out between the cylinder and the gun's breech that the other chambers would have gone off had they been loaded. A way to solve that problem, White decided, was to pack a piece of leather behind the gunpowder. A hole in the leather would convey burning fulminate from a percussion cap to the gunpowder. Again he prepared the borrowed gun for his next trip to the Phoenix basement. This time, he was satisfied with the result; there was no indication that a blast from one chamber would ignite others.

But the gun needed more work to make it a proper repeater, White reasoned, so he kept experimenting, adding various parts that he thought would result in a perfect firearm. By the middle of 1854, White was still making test trips to the basement with varying success. Now he was working not for his brother but directly for Colt, making revolver parts. White thought he had something worth patenting but held off applying. He considered showing his invention to Samuel Colt himself, he said later, but kept mum about it when his brothers told him that they would all lose their jobs if the boss learned that one of them had been experimenting with revolvers. They knew what they were talking about, they warned, because that had happened to other employees who tried to play inventor. White decided to bide his time.

White's work at the Colt factory ended in December 1854, when, he said later, Colt wanted his job performed by a direct employee, not someone contracting for it from the outside. Freed from Colt's control, White thought the moment ripe to offer his gun to the world, which meant applying for a patent.

On April 3, 1855, the US Patent Office gave him what he wanted: Patent No. 12,648 for "Improvement in Repeating Fire-arms." The gun that had finally completed its journey from White's mind to reality, at least on paper and in patent models, was a bit of an oddity. For one thing, White installed "a stout metal plate" covering the front of the cylinder in case neighboring chambers ignited at the same time, a disaster called "chain fire." That was to stop errant bullets from tearing apart the action, though it seemed to pose as much danger as it was supposed to prevent. One part of the patent held promise: the bored-through cylinder. But White knew—in fact, everyone knew—that a revolving cylinder that locked up in line with the barrel when the gun's hammer was pulled back was Samuel Colt's alone to produce. Anyone wanting to make money from bored-through revolver cylinders without having to face Ned Dickerson in court needed to wait until Colt's patent expired in February 1857—and hope that Colt wouldn't get his patent extended.

By all accounts Samuel Colt's London factory should have made him even richer. His was the first American armory overseas in an era when the Old World began to recognize United States excellence in industrial innovation, and who better to drive that success than the driven Samuel Colt? Even better, his gun factory on the banks of the River Thames had geared up just as its host country prepared for war and needed weapons. But London turned out to be a losing proposition.

There were several possible culprits, including competition from a British revolver firm the host country started to favor and mismanagement by James Colt, one of the inventor's brothers, whose job was to run the London enterprise. There was even a suggestion that James may have skimmed off some of the profits. Whatever happened, frayed filial bonds were strained so severely that the two brothers ended up squaring off in court over money. The rift never healed. And in the fall of 1856, Colt's London factory stopped making guns.

From then on the Connecticut factory would have to be the source of Colt guns for whoever in the world wanted them. Planning for a new building began the year after Colt's Crystal Palace triumph and his lecture to the British Institution of Civil Engineers. First he bought 250 flat acres everyone else thought suitable only for grazing, because each spring melting snows sent the Connecticut River flooding over its banks and across the low land. Colt solved that problem by building a dike to keep the river within bounds.

The new home for Samuel Colt's revolver business would become the largest private armory in the world. Atop its blue onion-shaped dome flecked with golden stars would rear the statue of a horse, called the "rampant colt," symbol of the man behind it all, the inventor and businessman whose name had become synonymous with his revolver. Colt was now a monarch of sorts, "Emperor of the South Meadows," as *The Hartford Daily Courant* wryly called him, holding sway over Coltsville, the two-hundred-acre complex where his workers and their families lived, prayed, danced, and spent their time when not engaged in producing guns for the world. On a hill above Colt's industrial fiefdom rose a mansion of baronial splendor. Soon called Armsmear, it would be Samuel Colt's home for the rest of his life.

With his revolving-cylinder patent closing in on its expiration date and other gunmakers eager to dive into the market, it might seem odd for Colt to invest so heavily in a massive expansion. But spending for him was a longstanding habit. Besides, he had plenty of eager customers in the United

States and abroad and was lobbying hard in Washington to have his patent extended once again. Colt was also looking for new developments in technology that would keep pace with the market if not move ahead of it. One of these was the cartridge. Experiments in Europe held promise, but not enough yet for full-scale production in the United States, as far as Colt was concerned. Once a worthy cartridge became available, he wanted to be ready. For that he would need a bored-through cylinder.

Whether or not Colt ever knew about Rollin White's secret experiments in his Hartford factory, he did learn of White's patent. Apparently he was interested in only one feature: the bored-through cylinder. In a letter late in 1855, Colt's patent attorney J. N. McIntire wrote to his client, "White has 5 patents and many claims, but the only claim covering any *broad* feature is *1st claim* of pat. No. 12,648. dated apl. 3rd. 1855." This claim covered "the construction of a rotating cylinder with chambers extending 'right through.'" McIntire told Colt that he thought the French had come up with something similar before 1855, and that he would look into it. Charles M. Keller, a prominent patent attorney who had been the first person to become an official patent examiner, told Colt that a French inventor's patents did, indeed, invalidate White's. "I am of the considered opinion," Keller said in a letter to Colt, "that Mr. White's claims can be set aside if suit is brought." Nevertheless, Colt took no action either to challenge White's patent or to license it.

But another gun entrepreneur did take action. Daniel Baird Wesson also learned of White's patent, and while he thought the strange gun that had emerged from White's imagination was worthless, its bored-through cylinder was not. That feature alone could be the key to success for the fledgling Smith & Wesson partnership. Not only the feature, but the patent that came with it. Stuffing leather into the back end of a cylinder's chamber to hold in powder had no appeal for Wesson, because he had been working on a better way to use an open chamber: the completely self-contained metal cartridge.

The prospect of a deal with White was a delicious one for the partners, especially Daniel Wesson, and not just because it promised financial rewards. It was also that he could turn the tables on powerful Samuel Colt. Five years earlier, in court, Colt had forced Edwin Wesson's revolver off the market and deprived Daniel of potential profits by claiming patent infringement. Now, if he could lock up Rollin White's patent, Wesson would have his name on a gun that Colt dared not copy unless he himself wanted to be a defendant in another patent fight, one that even the resourceful Ned Dickerson would lose.

On October 31, 1856, less than four months before Colt's revolver patent was set to expire, Wesson reached out to White.

"I notice in a patent granted to you under date of Apr 3 1855 one claim," Wesson wrote in a letter brimming with careful precision, "extending the chambers of the rotating cylinder right through the rear end of said cylinder so as to enable the said chambers to be charged from the rear end either by hand or by means of a sliding charger operating substantially as described." Then Wesson got to the point. That feature, he said, "I should like to make arrangements with you to use in the manufacture of firearms. My object in this letter is to enquire if such an arrangement could be made and if so on what terms."

It was now up to Rollin White to bargain for what must have seemed a gift from heaven.

19

THE SMITH & WESSON MOMENT

Rollin White wasted no time in answering Daniel Wesson's letter. The day he received it, White sent word that he was willing to meet Wesson and discuss terms. They set a date: November 17, 1856.

This was the moment White had been waiting for. No one had leapt at the chance to put his patent to use, though word of it had been out there in the usual places, including *Scientific American,* for more than a year and a half. Now with a wife and three-year-old daughter to support, White had returned to the retail selling he had abandoned to work for Colt and invent guns. Two days before Wesson wrote to him about his revolver patent, an advertisement appeared in *The Hartford Daily Courant* announcing that Rollin White & Co. was open for business, ready to provide fashionable clothing in new styles, "PRINTED FABRICS, of every price," as well as "FRENCH PRINTS, SHAWLS, FRENCH FLANNELS, BAY STATE PLAIDS, IRISH POPLINS, &c., &c." Now that Wesson had dangled before him the prospect of mass-producing a gun with his invention as a key component, White's world was opening up.

The meeting went to everyone's satisfaction. The document they all signed (Horace Smith was there, too) gave the gunmakers the exclusive right to use White's open-ended cylinder; the inventor retained the right to approve subcontracts. The next day, White was paid $497.50 with a promise that he would receive twenty-five cents for each revolver Smith & Wesson produced, once the Colt cylinder patent was history. All White had to do was sit back and enjoy the money that would roll in. There was one catch, hinging on the example set by Wesson's deceased brother. Edwin Wesson had instructed Daniel in two significant ways. First was how to make fine firearms as elegant as artwork, as finely machined as skilled hands could craft. But by his tragic example, Edwin had also taught Daniel what not to do in business and to be very careful about patents.

The memory of his older brother chasing down infringers of Alvan Clark's muzzle patent was strong in Daniel. Wesson insisted that White be required to "defend against and defray the expenses of any infringement" of his patent. The partnership was to license the patent, not own it. That meant that if someone decided to make a gun using White's patent, it would be up to him, not Smith & Wesson, to spend his own money going after the infringer. Under the terms of the agreement, White had no choice; he would have to go after them. Wesson knew how painful and costly that could be, but White was eager to move forward and put his signature on the contract. What made the situation sweeter for Wesson was that White had come up with his idea while working for Wesson's courtroom nemesis, Samuel Colt. And that the idea had become reality with revolver parts White had taken from Colt's factory. If all went as Wesson hoped, it could mean that it was Smith & Wesson's turn to be aggressors in future patent wars at someone else's expense.

Five hundred dollars went a long way in 1856, but once it was gone, White would have no more gun money coming until Smith & Wesson sold revolvers. This they wouldn't do until they got production up and running. And, of course, Colt's patent was still an obstacle, at least for a couple of more months. Better to stick with clothing in the meantime, so Rollin White & Co. stayed in business.

The day they signed their contract with White, the partners opened a fresh set of books to launch the Smith & Wesson Revolver Factory of Springfield, Massachusetts, just a few miles north of Hartford. Wesson, who was still living in New Haven but would move to Springfield in April, made the larger contribution to the startup: $2,003.63. Horace Smith's share was $1,646.68. By the end of January 1857, they had a metal model of what they hoped would be their first production revolver. The shop where Horace Smith and Daniel Wesson started making revolvers to compete with Samuel Colt's empire was no multi-story, city-dominating armory. Their factory, if one could call it that, was a leased floor in a stove- and pipe-making firm on Market Street in downtown Springfield. Below them William L. Wilcox's main business may have been stoves for cooking and heating, but he also offered a variety of metal products made to order, from furnaces to garden chairs to hat and umbrella stands made of iron.

In late March 1857—one month after Colt's revolver patent expired, despite his intense lobbying for another extension—Smith & Wesson announced that soon they would be making revolvers whose cylinders loaded from the rear with self-contained cartridges. "They will put in an engine of twelve horse power, and probably employ about twenty men," announced *The Springfield Republican* on March 26 in a column titled "New Operations in Business and Property." The partners, the paper said, "calculate to get under headway in a month."

The new guns' copper-cased cartridges "are water tight, and are slipped into their locations with great ease and rapidity," the *Republican* went on to report, adding that the weapon was light and immune from "all danger of a double discharge of barrels." Wrapping up its praise of the town's new product, the paper told its readers, "The pistol is very beautiful—in fact, one of the very neatest and prettiest revolvers we have ever seen."

This first Smith & Wesson revolver certainly was petite—smaller even than Colt's 1849 pocket pistol. There was no flamboyant engraving around its cylinder, which held seven rounds of low-powered .22 caliber ammunition. The partners called it the Model 1. Smith & Wesson advertised their new gun as "the seven shooter," proclaiming that it was "the lightest Revolver in the world, and shoots with as much force as any other arm." The first claim might have been true; the second was not. Mark Twain, who turned his wit on Ethan Allen's pepperbox pistol, also took aim at the Model 1:

> I was armed to the teeth with a pitiful little Smith & Wesson's seven-shooter, which carried a ball like a homoeopathic pill, and it took the whole seven to make a dose for an adult. But I thought it was grand. It appeared to me to be a dangerous weapon. It only had one fault—you could not hit anything with it. One of our "conductors" practiced awhile on a cow with it, and as long as she stood still and behaved herself she was safe; but as soon as she went to moving about, and he got to shooting at other things, she came to grief.

Perhaps the revolver's small-caliber bullets didn't pack much of a punch, but landing in quantity at their destination helped make up for what they lacked in potency. The gun was also easily concealed.

Rollin White had kept his clothing shop going, and thank goodness. Despite the early praise in *The Springfield Republican*, Smith & Wesson was turning out hardly any guns. In fact, by the end of 1857, the firm had made a total of four revolvers, which meant that White received only one measly dollar for the use of his gun patent. Such was the price of a start-up business with a new product. Holding out for a better year in 1858, White depended on sales from his shop on Hartford's Main Street, where he offered dresses for all occasions: "for breakfast, and dinner, and balls," as the advertisements said, and dresses "to dance in, and flirt in, and talk in," even dresses "in which to do nothing at all." The unimpressive year in gun production for Smith & Wesson ended with a wider bust in the Panic of 1857.

Until then the 1850s had been a prosperous decade for the United States. The country's economy had expanded, fueled partly by California gold, and migration west was in full swing. When the gold supply started to drop, high-flying investments seemed riskier than they had been a few months earlier, which in fact they were. Bank loans, especially to railroads, turned out to be more speculative than bankers had thought. When word that the Ohio Life Insurance and Trust Company had suspended payments on its notes spread rapidly through the financial world (thanks to Morse's telegraph), the panic began. As other banks refused to honor their notes, credit dried up. Hartford was not spared. In October a number of merchants there, including Rollin

White & Co., announced that they would still accept payment for their goods in notes from banks that refused to exchange hard currency for them. Smith & Wesson survived into 1858, apparently with little problem. Rollin White & Co. did not. By the end of the first week of January, White had left the store on Main Street, its contents to be sold off to close the business. By April he and his family had moved to Iowa, where his younger brother Mason then lived.

Samuel Colt also survived the Panic of 1857—comfortably so. Westward migration had peaked in the years before the Panic, but Easterners, including recent arrivals from Europe, continued to cross the continent in search of better lives. And, as always, they needed guns, especially Colts.

The approaching expiration of his revolving-cylinder patent didn't curtail Colt's spending. After a long honeymoon in Europe—he married Elizabeth Jarvis, daughter of an Episcopal minister, in June 1856—he and his bride settled in at Armsmear. During his sojourn abroad, Colt had ordered the best furnishings from the fashionable firm of Ringuet-LePrince & Marcotte of New York and Paris, choosing some of the fixtures himself. French furniture was chic among at least some other wealthy Marcotte clients. Naturally Colt tended to be showy, but in a few of Armsmear's rooms he stepped back from excessive opulence. Selections for his library and dining room, for example, were "robust, sturdy, and conservative."

After deliveries started coming to Armsmear in late January 1857, Leon Marcotte thought it was time for Colt to begin paying for them. Colt thought otherwise. He wanted to wait until everything was finished. Then he complained about the charges. Back when he was a struggling inventor and laughing-gas entertainer, Colt may have had an excuse for dodging bills, as he did with Baltimore gunsmith John Pearson. But in the 1850s, when he was a man of serious means, he continued the practice.

"I have been disappointed that you have not taken notice of my demand of money, as we want it just now after such a great advance of funds," Marcotte wrote in a letter to Colt. "We would not ask you to send us the whole amount of our bill, but half of it would be of great convenience now."

Having heard nothing from his client, Marcotte wrote again. "Hard pressed for want of money I started for Hartford . . . with the expectation to meet you & get something on account of our bill but was very much disappointed at not finding you at home. Please send us by return mail $5,000 or $6,000 or more if you can, as we want cash very badly, you will greatly oblige."

This time Colt got back to him. "I have just returned home & find your urgent letter for an advance of $5,000 upon your bill for furniture. I enclose my check for the amount to accomodate [*sic*] you even though it is directly against my rules to pay or advance money for account of contract until all business relating to it is completed to my entire satisfaction." That was not the way its business was usually carried on, Marcotte advised. For the future Colt and Marcotte agreed to complete orders in stages with payment for one part to come before the next part began.

Colt's armory was pulling in a steady stream of spending money as guns flowed out. Springfield Armory had been a training ground for skilled mechanics as well as a laboratory for industrial innovation. So too was Robbins & Lawrence in Vermont. Now Colt's factory was in much the same role. Another inventor who would make his mark in repeating firearms had been learning and inventing under the Colt factory's blue dome. In three years Christopher Miner Spencer would attach his own name to a gun that would help restore the Union.

20

OF SILK AND STEEL

Christopher Spencer could thank grandfather Josiah Hollister for fueling his natural interest in guns, but he also owed a debt to Frank Cheney's improved Rixford Roller. The roller, which had nothing to do with weaponry but everything to do with Spencer's future, revolutionized silk manufacturing and helped save the Cheney silk mill where Spencer got his first real job.

In the early nineteenth century much of the country, including Connecticut, went insane over silk. Plant a few mulberry trees, whose leaves silkworms feasted on, let the creatures spin cocoons, then unravel the cocoons, and, voilà, you had abundant silk to sell. Or so the thinking went. And it worked, for a while. Prices skyrocketed for imported white mulberry trees—they were better for the purpose than native red mulberries—which were planted wherever they would grow, and that was pretty much everywhere. Silk production was popular elsewhere, but by 1840 Connecticut's yield was three times that of any other state. Then the bubble burst.

First came the Panic of 1837 that rocked businesses nationwide and helped drive Samuel Colt's first firearms venture into bankruptcy. Then came the collapse of the market in mulberry trees. Then in 1844 blight hit them. Even healthy white mulberries plummeted in value. Fortunes were lost; businesses closed. The Cheneys had charged full force into the silk business, planting nurseries not only in Manchester but also on leased land in New Jersey and on a farm in Ohio. The Connecticut mill closed for a while but was resurrected by a pair of Cheney brothers, both successful portrait painters, from their own finances.

That the Cheneys were back in business was good news for Christopher Spencer, whose uncle Owen had a good friend in Frank Cheney, the family's most invention-prone member and a hands-on manager. An introduction was made, and young Spencer joined the Cheney workforce as an apprentice. Immediately Frank Cheney and the boy established a bond that would push "Crit" into a career path.

Spencer wrote of Frank Cheney late in his life:

> I remember him in the little old mill by the brookside with its machinery turned by the water-wheel which he so laboriously fitted up, going about among the winding machines with words of encouragement and instructions for all the little company of girls and boys then engaged in the operation of reeling, winding and spinning. The deft tying of knots, picking up of threads from the tangled skeins, all are pleasurable recollections of my first Summer in the Mill, at the age of 14, when I first became imbued with the idea of becoming a mechanic.

Frank Cheney—"the ruling spirit in the work of the factory," in Spencer's view—did more than encourage young workers. When something needed repair, he often did the job himself, even wading into the stream to repair the mill's waterwheel on which the entire operation depended. At times Frank took duty as the night watchman and cleaned the lamps. For Spencer the mill was a healthy place where managers and laborers toiled together.

Spencer's association with the Cheneys and his timing were fortunate. In 1847, the year Spencer signed on, Frank Cheney came up with an idea to make silk more profitable again. Doubling, winding, and twisting the fine silk threads was a tedious, labor-intensive process. Using devices developed for cotton spinning, Frank patented a machine that did the work more efficiently and more quickly than the so-called Rixford Roller, which Cheney employee Nathan Rixford had made. Costs dropped and profits rose.

Spencer's first employment with the Cheneys—there would be several—lasted only a year. In 1848 he became apprenticed to a Manchester machinist named Samuel Loomis, apparently at a more advanced level than he had enjoyed at the Cheney mill. Inspired by reading *Comstock's Common School Philosophy,* which discussed the science of mechanical things, Spencer made his first machine: a stationary steam engine.

That winter he took a break to attend a prep school in Wilbraham, Massachusetts, the only formal schooling the boy would have. Then in the spring of 1849, he was back in Loomis's shop for a few months before returning to the Cheney mill, this time as a journeyman machinist. While Spencer was developing his skill with machines, his mentor Frank was stricken with gold fever like so many others and had joined the boy's uncle Owen on a voyage to California in search of riches. They would be home eighteen months later without gold but having established a lumber mill near San Francisco.

Encouraged by Frank, Spencer left the Cheney mill again to spend six months making machinists' tools in Rochester, New York, followed by another six months at a locomotive repair shop, and then some time at a Massachusetts armory. In 1854 Spencer returned to Connecticut to work at Samuel Colt's armory, helping to build and improve gunmaking machinery. He would stay with Colt for about a year before the Cheneys lured him back.

By that time Spencer's bouncing between various businesses where machines were crucial had given him broad mechanical knowledge. At the Cheney mill once again, he became superintendent of the machine shop in the family's Hartford operation. And he kept tinkering. Although now a man, Spencer remained buoyantly youthful, always trying to figure out how something could be improved, never hobbled by melancholy. Mechanical problems did not keep him awake at night. "I go to sleep thinking about them," he told a friend later in life, "and often in the morning I have the solution."

The sin of slavery festered in the nation more than ever as 1860 approached. Sectional tensions were rising. Abolitionists, largely in the North, called for an end to a system in which one human being could own another.

Southerners, whose economy depended on slave labor, resisted any thought that the millions of black men, women, and children they held in bondage would ever be free. Efforts to appease both sides by permitting some territories in the West to decide whether they would allow slavery were proving unsatisfactory. The Fugitive Slave Act of 1850, which required that escaped slaves caught in free states be returned to their masters, caused increased friction in the North, where fewer citizens were willing to tolerate the inhumanity of the "peculiar institution."

The Cheneys detested slavery. Frank's brother Charles, for example, was "a damned Abolitionist," to use the term Charles's son, also named Frank, heard schoolmates hurl when he and his family lived for a few years in Mount Healthy, Ohio. Ohio had abolished slavery in 1802, but anti-black sentiment remained deep enough in many citizens to make it dangerous to harbor slaves fleeing north. Young Frank wondered about the transient visitors who came to the Cheney farm after dark and left before daybreak—black men, women, and children whose arrivals were signaled by mysterious knocks on a window in a sequence the family recognized. After repeated pestering, Charles decided to trust his son with a secret. These were runaway slaves who needed a safe place to stay on their journey to freedom, and the Cheney farm was the first station of the Underground Railroad out of Cincinnati.

"Nothing ever came to me which brought such a feeling of individual responsibility as did the knowledge of this secret," young Frank recalled years later, "which my father thought it was best and safest to impart to me, though I was only a boy ten or twelve years old."

The boy slept in his father's room and did not always hear the telltale knocks, but sometimes they woke him with a start. In whispered voices the new arrivals and "the conductor" who brought them gave Charles information he needed about their guests for the night and how to get them on course again in the morning. If the Cheneys feared slave catchers were in close pursuit, they fed the travelers and sent them quickly on their way. Once Charles kept a man and wife with their three children out of sight for several days, "as it was known that we were under suspicion and the slave-hunters were on their track."

It was while he was working for the Cheneys in 1857 that Christopher Spencer began thinking about how to invent a repeating rifle that held its cartridges in the butt stock, not under the barrel. This was a tricky proposition, requiring more movements and mechanical complexity than feeding cartridges from the front. Yet that's what Spencer wanted to do from the beginning. He didn't say why he opted for a butt-stock magazine, but he may have wanted to avoid a patent fight with Oliver Winchester, whose Volcanic repeater used the under-the-barrel method.

Over the next couple of years, Spencer spent his spare time—which was scant, given his sixty-six-hour workweeks—experimenting to come up with the right system. When he was far enough along, he made a wooden model. Then he sought help from Richard S. Lawrence, who had left Vermont and was now superintendent of the Sharps Rifle Manufacturing Company in Hartford making breechloaders. Lawrence sold Sharps parts to Spencer, who designed some of his own mechanism around them. With financial support from his successful wool-merchant father, Spencer made a prototype and prepared to apply for a patent. He also made his employers happy by doing some inventing for them, which resulted in his first two patents: one for a machine that labeled spools of thread and the other for a machine that did the spooling.

Spencer was finally satisfied with his new gun. While the mechanism was intricate and complex, making it work was not. Pull down on the long, looped trigger guard, and the action swung open, ejecting a spent cartridge casing and allowing the next cartridge to be fed by a spring from a tube in the buttstock into a chamber at the rear of the barrel. Pull up on the lever, and the action shut tight. Then all the rifleman had to do was cock the hammer and shoot. After its seven shots had been fired, the empty tube that had contained the cartridges was pulled out and filled with seven more cartridges before being shoved back into the gun for more shooting. In the hands of trained soldiers, this weapon could be devastating.

On March 6, 1860, less than five months after receiving his first patent, Christopher Spencer was awarded his third, not for a piece of factory machinery this time but for his new weapon. As with all creations, a patent takes the inventor only so far. Unless Spencer could make the gun and find a market, all the patent gave him was the right to call the idea his own. Frank Cheney, cheering him on, had faith in Spencer's invention. If war was to come, perhaps this new, fast-firing weapon could be useful to the government. The Cheneys had money and well-placed friends. Perhaps they could help Spencer again.

21

A GATHERING RUSH FOR WEAPONRY

On the afternoon of March 5, 1860, the day before the US Patent Office awarded Christopher Spencer exclusive rights to his design for a repeating rifle, a train carrying a passenger scheduled to give a speech at City Hall that night rolled into the Hartford station. Abraham Lincoln had been in New Hampshire to visit a son at school and give him needed encouragement in his studies. The trip's public purpose was campaigning for the presidency of a nation on the verge of ripping itself apart.

Strife between North and South, between free and slave states, had been intensifying for years, fed partly by the vast new lands the country had conquered in the Mexican-American War. Would the new Kansas and Nebraska territories allow slavery within their borders? In 1854 Congress voted to let the people there decide for themselves, a move that led to an influx of both pro- and antislavery settlers and so much violence that the term "Bleeding Kansas" was born. One of the men who shed blood there was abolitionist John Brown, who went on to attack the federal armory in Harpers Ferry, Virginia, in October 1859 in the hope of seizing weapons he would turn over to slaves who would then revolt against those keeping them in bondage. Two of his sons were killed in the raid, Brown was captured, and there was no slave revolt, though Brown and his forces managed to hold the armory for a short time. They would hold the nation's attention far longer.

The attack's short-term goal failed, but it helped further polarize the nation over slavery. Southern fears of Northern antislavery programs ratcheted up even more, while many Northerners saw in Brown a martyr for the noble struggle against the evils of enslavement. Six days after Brown was sentenced to hang, Ralph Waldo Emerson called the condemned man "the new saint awaiting his martyrdom, and who, if he shall suffer, will make the gallows glorious like the cross." When he left his jail cell to be hanged on December 2, Brown slipped a note to a guard. On it, he had written a prophecy: "I, John Brown, am now quite certain that the crimes of this guilty land will never be purged away, but with blood."

Three months later, as Abraham Lincoln began his New England speaking tour, presidential politics increasingly drew the nation's attention, with slavery a burning issue. The Republican front-runner in the waning winter of early 1860 was staunch abolitionist New York senator William H. Seward, but Lincoln was coming on, buoyed by a speech he had given the week before at New York City's Cooper Union. In that address the lanky rail-splitter from Illinois had stirred those who heard what he had to say and those who read about it. Whether slavery would be allowed in the territories was a subject Lincoln addressed head-on. In his Cooper Union speech, with eloquent strength, Lincoln declared it must not be. At the end, his voice rising, he called on the throng before him to fight the spread of slavery: "Let us have faith that right makes might, and in that faith, let us, to the end, dare to do our duty as we understand it!" The crowd loved those words—and the man who spoke them.

Now on his one-day stop in Hartford, the travel-weary Lincoln was ready for his next speech. With time to spend before he was to give it, he went to the Brown & Gross bookstore opposite the State House in the heart of town. There Lincoln met patrician newspaper editor Gideon Welles, a Republican national committeeman and an ex-Democrat who had worked against Seward, someone Welles didn't care for. The two men sat together in front of Brown & Gross on a bench that the bookstore proprietors had put there for people awaiting carriage rides. What they talked about is not recorded, but they sat there long enough to draw notice. Welles, prominent in town and beyond, was easy to spot, with his voluminous white beard and ill-fitting wig of a different hue sitting atop his large head. Passersby stopped to chat, wondering whether the tall man with Welles was the giant killer from Illinois they had read about in the papers. Neither man knew it at the time, but both would have an outsized impact on what was to come: Lincoln on the country's future, Welles on Christopher Spencer's.

That night, a "pressed down, shaken together and running over" crowd in City Hall gathered to hear what the gaunt and homely former frontiersman had to say. Hundreds more couldn't get within hearing distance. Lincoln arrived in the mayor's carriage in which his afternoon companion at the bookstore had declined the offer of a seat; Welles preferred to walk with friends and political allies in a torchlight procession to City Hall, a mighty Doric temple with large porticos at each end. Once in the third-floor hall where the event was to be held, however, Welles took a seat on the dais, where he could see and hear the speaker up close. Lincoln did not disappoint. He modeled his speech on the one he gave at Cooper Union, including his resounding "right makes might" call to action at the end, which brought the crowd, including Welles, to its feet, with roars of approval.

After Lincoln finished, he was escorted from the hall by a cornet band and more torchbearers, young men in militaristic formation, who were part of a movement called "Wide Awakes," whose goal was to protect Republican marchers and speakers from Democratic assaults. This night was one of the Wide Awakes' first assignments—they had come into existence only a few days before—their distinctive glazed black caps and capes shielding them from oil that dripped from their torches. The movement would soon spread from Hartford throughout the North, with thousands of young men parading at night, pushing hard for Republican victories.

Late the next afternoon, a raw and gusty March 6, Lincoln was back on the train, heading to New Haven, where he was to give another speech that evening. But before he left, he had two missions. One was to tour the Colt factory on the banks of the Connecticut River on the east side of town, and the Sharps rifle factory on the appropriately named Rifle Avenue on the west. Lincoln's second mission was to sit down again with Gideon Welles, this time just the two of them in the privacy of Welles's office at *The Hartford Evening Press*.

Lincoln impressed Welles, who found him "in every way large, brain included, but his countenance shows intellect, generosity, great good nature, and keen discrimination. . . . He is an effective speaker, because he is earnest, strong, honest, simple in style, and clear as crystal in his logic." Welles was a force, a man of influence and opinion. With Abraham Lincoln on the rise, Welles—and his prestige—might rise too and with him the fortunes of Christopher Spencer. Among Welles's intimate friends and neighbors was Charles Cheney, who, like his brother Frank, supported Spencer's newly patented, fast-shooting firearm. If war was to come, the Army might use such a weapon. Although there was no factory to do the job and no funds to build one, the Cheneys were well positioned to see the Spencer rifle put into production when the time was right—and to make a profit. If Gideon Welles found a place in a Lincoln administration, his friendship could be useful. Perhaps then Christopher Spencer's relentless inventiveness might have more impact than merely increasing the efficiency of silk factories. It could equip an army.

When Lincoln visited Hartford, the Spencer rifle was nothing more than a dream reduced to paper accompanied by a model. The guns coming from the Colt and Sharps factories, on the other hand, were that and much more. Colt's workforce had swelled from 150 employees making 9,000 pistols in 1850 to 369 employees turning out 45,000 guns plus machinery for sale in 1860. The Sharps factory, once associated with Robbins & Lawrence, didn't exist in 1850, but ten years later its 300 workers built 10,000 breech-loading rifles invented by Christian Sharps, who had worked closely with the Vermont firm to develop and make his guns in the early 1850s. The world of American arms making had come a long way since Samuel Colt launched his first revolver business in 1836. The old system of small makers (Eli Whitney was an exception) who contracted with the national armories they depended on was giving way to larger private concerns whose owners set themselves on a path to personal wealth. Private capital launched these businesses, and while they courted government contracts, they struck out on their own with designs they or those they employed came up with. They were a new breed who defended patents jealously, rushing into court to beat back the competition. Among them were builders of business empires.

Colt led the way. He had tried to get his revolving-cylinder patent extended even after it had expired by wooing powerful people he thought could do him some good, including legislators he favored with attention and gifts. Colt lost that battle, but his production of revolvers kept going strong. In 1860 Colt added a new revolver to his line, bigger than the Navy, smaller than the Dragoon, perfect for carrying in a hip holster. It would be called the 1860 Army and begin coming off his assembly line just in time for serious orders from a military engaged in all-out war. And why wouldn't Samuel Colt be a success even without his exclusive patent? He was now more than a man; he was a brand. More than a brand, he was a gun, his name increasingly synonymous with a revolving pistol. People moving west were encouraged to follow this advice from an 1859 handbook for cross-country travelers:

> Every man who goes into the Indian country should be armed with a rifle and revolver, and he should never, either in camp or out of it, lose sight of them. When not on the march, they should be placed in such a position that they can be seized at an instant's warning; and when moving about outside the camp, the revolver should invariably be worn in the belt, as the person does not know at what moment he may have use for it.

What kind of revolver would be best suited to the journey? The handbook had advice on that too: "Colt's revolving pistol is very generally admitted, both in Europe and America, to be the most efficient arm of its kind known at the present day."

Things were going pretty well for Samuel Colt in 1860 except his health. In his mid-forties, he kept up his habit of drinking liquor, often the finest, and enjoying smoke he sent billowing from cigars he bought in abundance.

Then there was gout, which caused him pain and limited his movements. Colt kept moving, though, as he always did.

Other major gunmakers were also gearing up as the 1850s drew to a close.

Smith & Wesson had an obvious advantage when they made Rollin White's patented cylinder a feature of their first revolver. If a multichambered gun that reliably shot self-contained metal cartridges attracted a good chunk of the buying public, the partners could corner the market, for theirs was the only legal one available. And the diminutive size of Smith & Wesson's Model 1 made it attractive to urban dwellers and others who preferred to arm themselves discreetly.

In the first three years after they started producing their dainty revolver, Horace Smith and Daniel Baird Wesson had made six design changes. Altering a gun while it's in production was common in the private firearms industry at the time—less so in government armories—but the extent of Smith & Wesson's tinkering was unusual. By this time the partners had lots of experience making and designing guns, so it may be a bit of a surprise that there was so much need for improvement. Their gun was also hard to work on, thanks to the small side plate that served as a gateway to the weapon's interior mechanism. That problem, though a minor one, limited production. Nevertheless the partners were betting everything on their pocket-sized pistol and its wimpy though groundbreaking .22-caliber copper-cased cartridge. The bet was paying off. They produced more than eleven thousand Model 1s in the first three years, with enough back orders to justify moving their workforce out of the cramped space they rented above William Wilcox's stove factory.

In late 1859 they built what they needed: a four-story building on Stockbridge Street near the Connecticut River in downtown Springfield. It was an impressive structure, 150 feet long and 35 feet wide, occupying, with its outbuildings, a half-acre lot where orchard grass and apple trees had grown a few years earlier. For Smith and Wesson the completion of their new building was a perfect reason to celebrate. So on the last Friday of December they threw open its doors to host a grand ball, with supper served on

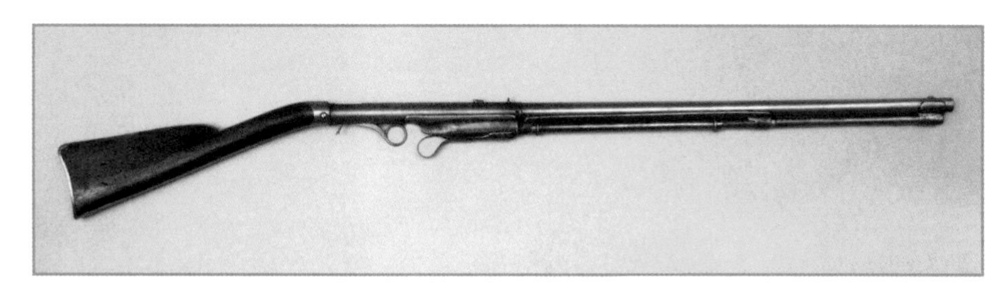

Walter Hunt's "Volition Repeater," a revolutionary but failed repeating firearm that was a precursor to rapid-fire weapons. *Courtesy of the Buffalo Bill Center of the West, Cody, Wyoming, MS020 Winchester Repeating Arms Co. Collection*

Showman, inventor, business titan, huckster Samuel Colt, holding an 1851 Navy Revolver, one of his company's most popular handguns.

Engraving by John Chester Buttre, circa 1855, Wikimedia Commons

Eliphalet Remington, founder of the firm that bore his name and produced guns for more than two centuries.

Edwin Wildman, Famous Leaders of Industry, 1921, Wikipedia Commons

Carpenter turned shirtmaker turned gunmaker Oliver Fisher Winchester.

Courtesy of the Buffalo Bill Center of the West, Cody, Wyoming, MS020 Winchester Repeating Arms Co. Collection

Horace Smith, who rose from apprentice bayonet forger to firearms industry titan.

Courtesy of the Lyman and Merrie Wood Museum of Springfield History, Springfield, Massachusetts

An Allen & Thurber pepperbox pistol like the ones Horace Smith helped build when he worked for that firm.

Courtesy of International Military Antiques, Inc., ima-usa.com

Daniel Baird Wesson, holding what appears to be a pepperbox pistol, standing beside his wife, Cynthia Hawes Wesson.

Courtesy of Roy G. Jinks, historian, Smith & Wesson Historical Foundation.

Christopher Miner Spencer, ebullient inventor of a repeating rifle used by the Union in the Civil War.

Courtesy of the Windsor Historical Society Windsor, Connecticut

Samuel Hamilton Walker, the Texas Ranger and US soldier who helped design Samuel Colt's first successful revolver.

Library of Congress, Prints and Photographs Division

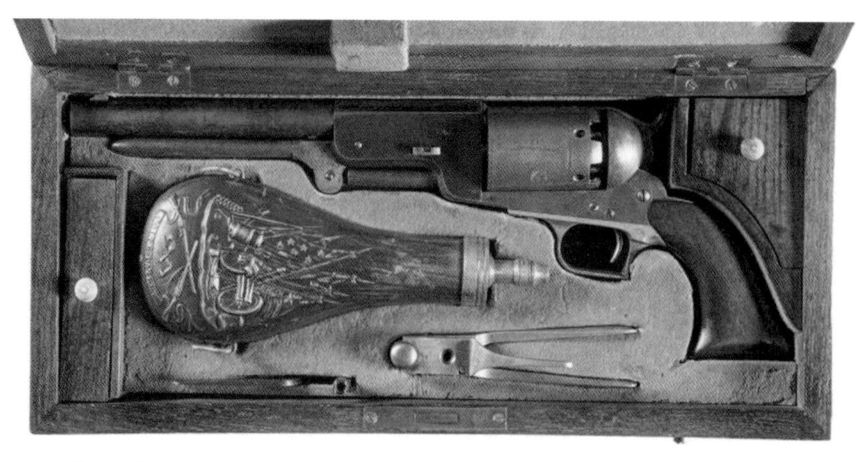

A cased "Walker Colt" with tools and powder flask, which sold at auction in April 2018 for $1.8 million.

Courtesy of the Rock Island Auction Company, Rock Island, Illinois

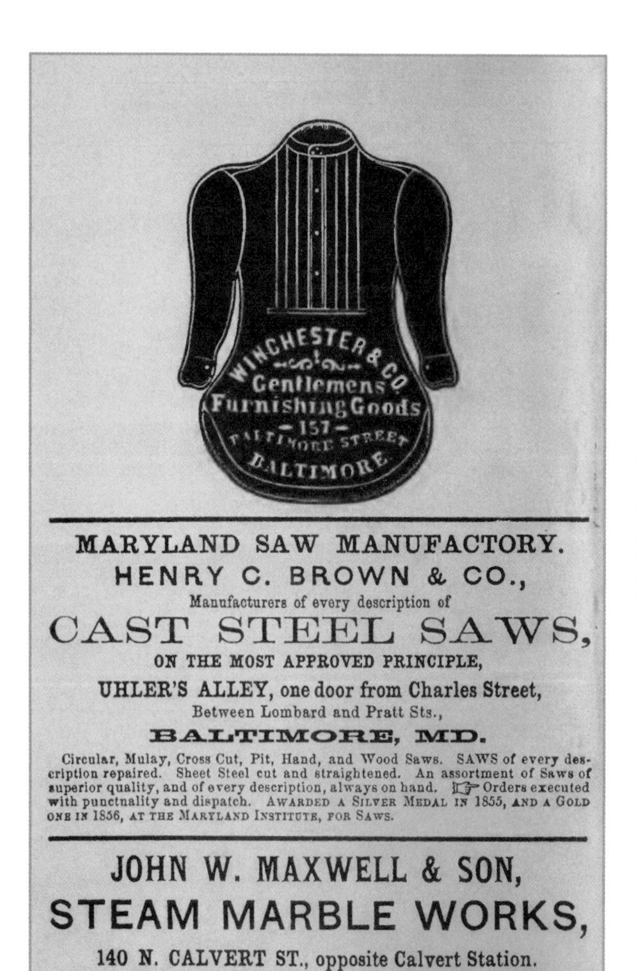

MARYLAND SAW MANUFACTORY.
HENRY C. BROWN & CO.,
Manufacturers of every description of
CAST STEEL SAWS,
ON THE MOST APPROVED PRINCIPLE,
UHLER'S ALLEY, one door from Charles Street,
Between Lombard and Pratt Sts.,
BALTIMORE, MD.

Circular, Mulay, Cross Cut, Pit, Hand, and Wood Saws. SAWS of every description repaired. Sheet Steel cut and straightened. An assortment of Saws of superior quality, and of every description, always on hand. ☞ Orders executed with punctuality and dispatch. AWARDED A SILVER MEDAL IN 1855, AND A GOLD ONE IN 1856, AT THE MARYLAND INSTITUTE, FOR SAWS.

JOHN W. MAXWELL & SON,
STEAM MARBLE WORKS,
140 N. CALVERT ST., opposite Calvert Station.

Advertisement for the Winchester clothing store in Baltimore, then run by Oliver's twin brother, Samuel.

E. M. Cross & Co.'s Baltimore City Business Directory, 1863–64

Colt's display at the Crystal Palace in London, 1851, where he kept a stash of brandy to serve potential customers. *The London Illustrated News, December 6, 1851*

Armsmear, Samuel Colt's mansion in Hartford. The name means "the meadow of arms." *The Homes of America, Martha J. Lamb, ed., 1879*

B. Tyler Henry, inventor of a repeating rifle that ensured Oliver Winchester's success as a gunmaker.

Courtesy of the Buffalo Bill Center of the West, Cody, Wyoming, MS020 Winchester Repeating Arms Co. Collection

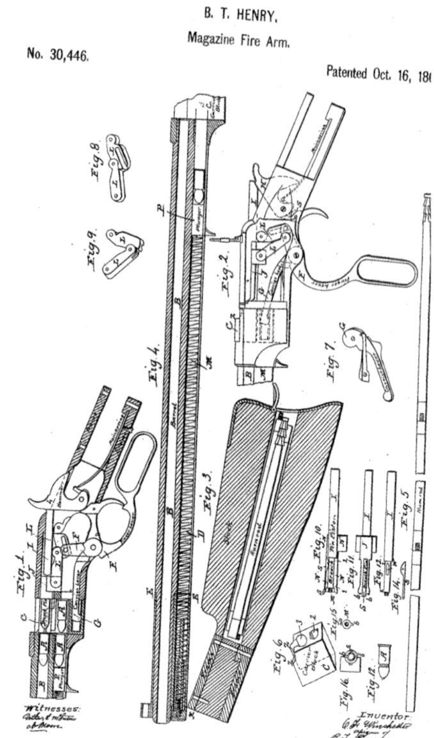

Patent drawing for the Henry rifle. Henry assigned the patent to Oliver Winchester. The two eventually had a falling out.

Public domain

Lushly engraved, silver-plated Henry rifle that Oliver Winchester gave to Navy Secretary Gideon Welles.

Courtesy of the Autry Museum, Los Angeles; 2012.2.18 (detail)

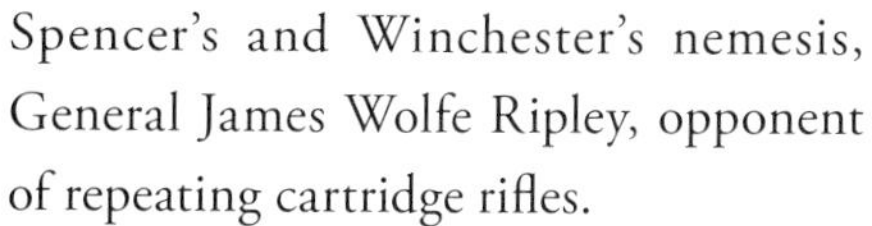

Spencer's and Winchester's nemesis, General James Wolfe Ripley, opponent of repeating cartridge rifles.

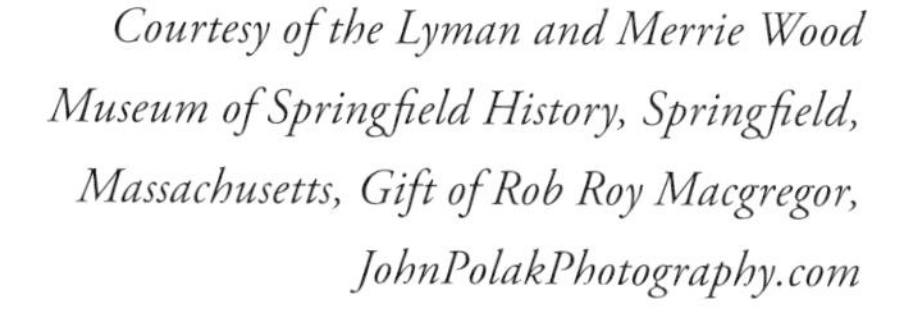

Courtesy of the Lyman and Merrie Wood Museum of Springfield History, Springfield, Massachusetts, Gift of Rob Roy Macgregor, JohnPolakPhotography.com

Col. John T. Wilder, whose Lightning Brigade used Spencer repeating rifles to devastating effect against greater forces in several Civil War battles.

Public domain

Custer directing the Michigan Wolverines, armed with Spencer rifles, in Pennsylvania just before the Battle of Gettysburg.

Detail from Custer at Hanover by Dale Gallon, www.gallon.com

President Ulysses S. Grant and Brazilian Emperor Dom Pedro II starting the Corliss steam engine at the 1876 Centennial Exhibition. *Wikimedia Commons*

The arm of what would become the Statue of Liberty on display at the Centennial Exhibition. Visitors stand atop the torch.

Library of Congress, Prints and Photographs Division

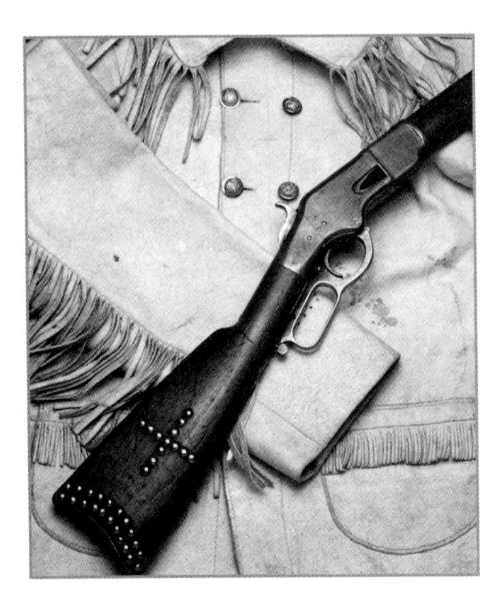

A Winchester 1866 carbine that purportedly belonged to Sitting Bull resting on Custer's jacket.

Courtesy of the National Museum of Natural History and National Museum of American History, Smithsonian Institution

Custer with an elk he shot while on the 1873 Yellowstone Expedition and his Remington rolling block hunting rifle.

Courtesy of the National Anthropological Archives, Smithsonian Institution, Photo Lot 166, William R. Pywell photographs from the Yellowstone Expedition the second floor and dancing to strains of the Hampden Quadrille Band on the floor above, where women's crinoline dresses had ample room to twirl about. Wesson and Smith invited their employees to attend, with the public also welcomed for a dollar a ticket.

When the new factory began turning out guns in January 1860, Smith & Wesson had four times the number of employees that had worked in its Market Street factory, including dozens of women making the special ammunition the revolvers required. The partners quickly filled the first and second floors with the most modern gunmaking machinery available, renting out part of the basement to Burbank Brothers, who made gold and silver thimbles and spectacles, and the third floor to a company that fabricated gold chains. "Capacious as are their new quarters," observed *The Springfield Republican* in an article celebrating Smith & Wesson's new building, "the room is already fully absorbed, and the time when they will 'cry for more' looms up in the near future."

The Republican's prediction was not far off. By the beginning of March, three months after the new building became operational, Smith & Wesson had received an order for five hundred pistols from South America, with other orders coming from London, as well as many western and southern US cities, to the tune of between two hundred and five hundred revolvers each month. With more orders than they could fill, Smith & Wesson had plans to add a new revolver to their so far single-product line, one larger and more powerful than the petite Model 1. This, they hoped, would find favor with soldiers if war came.

In 1858 Remington added to the product line a new item patented by a master mechanic working for the company, one intended to give gentlemen added protection when walking about: a cane that doubled as a rifle. For some time Europeans had been making weapons, including guns, hidden in canes, as had Allen & Thurber, but the Remington version was especially discreet because its entire single-shot mechanism was in the cane's upper shaft. An inconspicuous button served as the trigger. And of course the barrel was made of Remington cast steel. Advertised as "simple, safe and efficient," the

Remington gun cane was thinner than the typical weaponized walking stick, making it appear more like the ordinary thing a gentleman would carry with him on a stroll. For added panache a customer could have one with a handle in the shape of a dog's head. Gun canes were fine for male members of the urban elite, but real money had to come from products that appealed to a broader clientele, so Remington was also in the revolver market and would join Colt in arming the Union for the Civil War.

Oliver Winchester's shirts were selling well in 1860, but his New Haven Arms Company barely limped along. The company continued to trumpet the supposed excellence of its Volcanic repeater in advertisements and broadsheets with testimonials such as this from an ex-captain of a clipper ship in October 1859:

> I consider the Volcanic Repeating Pistol the *ne plus ultra* of Repeating or Revolving Arms, and far superior in many respects to Colt's much extolled Revolver. I have fired, myself, over 200 shots from it without even wiping the barrel—this is an advantage which no other Arm I know of possesses. I have had the Pistol with me at sea for more than eighteen months, on a voyage around the world, and find that, with the most common care, it will keep free from rust, far more so than Colt's.

The commander of another ship also waxed enthusiastic about Winchester's gun: "I have used a Volcanic Repeating Pistol for some months, on my last voyage to San Francisco, and in all that constitutes a good Pistol or Fire Arm, it has no equal, and excels all others I have ever seen in rapidity, efficiency and certainty of execution." But nothing topped the claim of an agency selling the guns in Illinois: "The Volcanic Repeating Pistol of [*sic*] Rifle is the greatest invention of the age."

A smattering of glowing testimonials for its product did not slow the New Haven Arms Company's likely journey to insolvency, the same fate its predecessor, the Volcanic Repeating Arms Company, suffered only three years earlier. The main problem was still the gun's weak ammunition; its caliber was large enough, but it was woefully underpowered. It took him a while, but shop superintendent B. Tyler Henry came up with a .44-caliber metal-cased cartridge with a lot more power. He also redesigned the Volcanic itself to accommodate the larger load. For the new gun he was granted Patent No. 30,446 on October 16, 1860. Henry immediately assigned the rights to his employer, Oliver Winchester. Despite their now obvious shortcomings and the prospect of a much better gun as soon as production would permit, the New Haven Arms Company kept insisting that Volcanic carbines and pistols were just what the modern shooter wanted. "These arms," claimed the company's Baltimore agent, who happened to be George Winchester, son of Oliver's twin brother Samuel, "were invented by several of the most ingenious mechanics of the country, who have spent years in bringing this most wonderful triumph of skill to perfection." While Oliver Winchester gathered enough funds to begin preparing to make a better rifle, nephew George hawked the troubled Volcanic as "the most powerful and most effective weapon ever invented, and will, doubtless, supersede nearly every Arm in the market."

Exactly three weeks after what would be called the Henry rifle received its patent, the nation went to the polls and elected Abraham Lincoln president. War was closer than ever. So was opportunity.

22

CALL TO ARMS, ANY ARMS

Samuel Colt was always war-ready, but the prospect of perhaps the richest market of all in the wake of Lincoln's election meant expanding his already massive factory. On December 20, South Carolina became the first state to secede from the Union. Mississippi and four others soon followed, to be joined by Texas in February, a month before Lincoln was sworn into office. Selling guns to states that had broken from the Union was wrong, according to Remington, but Colt continued sending his weaponry south for a while, as he had for some time.

"Run the armory night and day with double sets of hands until we get 5,000 or 10,000 ahead of each kind," he had instructed in a February 18, 1860, letter from Cuba, where he was resting and restoring his health. "I had rather have an accumulation of our arms than to have money lying idle, and we cannot have too many on hand to meet the exegencies [*sic*] of the time. . . ." For emphasis Colt closed with *"Make hay while the sun shines."*

Confederate cannon bombarding Fort Sumter in Charleston Harbor destroyed any lingering doubt about whether war would come. It had arrived. On that morning and into the afternoon of the next day—April 12 and 13, 1861—it was clear that there would be no turning back, not for the North, which wanted to bind the still-young nation together, not for the South, which wanted to preserve a way of life dependent on slavery.

It was a modest battle, as battles go, the only immediate casualty being a Union artillery private killed when a cannon discharged prematurely during a planned hundred-gun salute as the fort's flag was lowered in surrender. Daniel Hough, an Irish immigrant from County Tipperary, received the distinction of becoming the first man killed in America's Civil War.

The Union Army had enough guns for men already in service when the war began. According to the common wisdom at the time, the conflict wouldn't last long. At the very least the North needed to build up its military with volunteers, including the seventy-five thousand President Lincoln called for in the wake of Fort Sumter's loss, and the government quickly realized there weren't enough small arms to supply them. Output at the Springfield Armory was good and efficient. So was the case with the other national armory at Harpers Ferry, Virginia, until federal troops burned it on April 18, the day after Virginia voted to secede from the Union, to keep its stockpile of weapons out of Confederate hands. As it did when the Revolution began, the United States government initially turned to Europe for additional firepower, sometimes settling for guns that were not the best quality. The government also made contracts with private American gunmakers. Here was an opening for the new breed of firearms capitalists, and the Cheneys did not delay taking action.

Christopher Spencer was more inventor than businessman, but members of the Cheney family were both. And they had money plus connections. To get things moving on the new repeating rifle, the Cheney brothers made a deal with Spencer. From their own deep pockets, they would get production up and running and do everything they could to put it into the marketplace. They would also give the inventor royalties, while they retained the profits, if there were any. In return, the enthusiastic, ebullient Spencer would be the gun's primary salesman. As soon as the deal was signed, the Cheney brothers went into action by calling on friends in high places. Prominent among them was Connecticut Governor William A. Buckingham, a solid Union man and Lincoln loyalist who had introduced the future president when he spoke at Hartford City Hall.

"Permit me to introduce the Hon. Charles Cheney of this city who has some business with your Department," Buckingham wrote in a May 28, 1861, letter to Secretary of War Simon Cameron. "Mr. Cheney is a gentleman of high respectability and standing in this community and worthy of your confidence."

Seven days later more high-level pressure was brought to bear on behalf of the Cheneys. This time it came from old Cheney friend and neighbor Gideon Welles, whom Lincoln had made secretary of the navy three months before. In a letter to John A. Dahlgren, the recently appointed commandant of the Washington Navy Yard, Welles introduced his "special friends" the Cheneys, who had "a newly invented breech-loading arm, which they and others think superior to any thing invented; and I, with them, desire to have your opinion respecting it, or at all events, wish you to give it such examination as your time will permit." In case Dahlgren might take this instruction lightly, Welles closed with "I need not say to you, that my friends the Messrs. Cheney are among our best citizens, and that any attention to them I shall esteem a special favor."

Dahlgren might not have needed prodding from his boss. He knew weapons and was interested in innovations, so the chance to examine a radically new firearm would appeal to him. In fact Dahlgren himself invented a naval cannon named after him, and had developed a new friend and fellow mechanical enthusiast even higher up the political ladder than the secretary of the navy. That friend was Abraham Lincoln, who occasionally dropped by the Navy Yard to seek Dahlgren's advice about weaponry or just to pass time with him, even shooting in his company there on occasion.

Whether to please Welles or to satisfy his own curiosity, Dahlgren quickly put Spencer's gun to the test. Four days after the Navy secretary introduced the rifle-bearing Cheney brothers to him, Dahlgren wrote to Captain Andrew A. Harwood, chief of the Bureau of Ordnance and Hydrography, saying that the Spencer "operates so well that I am induced to bring it to your notice." Despite having a complicated mix of parts, he said, "the mechanism is compact and strong." Over two mornings, Dahlgren reported, the gun was fired five hundred times with only one failure, and that was probably caused by a faulty cartridge. "I can recommend that a number of these pieces be introduced for trial in service."

At this point there was no factory producing Spencer repeaters, only some working models emerging from the Cheney workrooms. If the Army in its desperation for weapons gave the Cheneys a big order for guns to be delivered quickly, they would face an impossible task. Their position was like Samuel Colt's when he agreed to make Walker revolvers for the Mexican-American War, or like Robbins & Lawrence when that firm snagged an order for ten thousand rifles: no factory, no workforce, no machinery, no materials. Nevertheless they pushed on, getting positive comments from people who could help.

If they were to win serious government contracts, the Cheneys had to work around the new Chief of Ordnance and former Springfield Armory superintendent James Wolfe Ripley. The crusty, stubborn Ripley was admired for his organizational skills and rock-ribbed devotion to duty. He had been out of the country inspecting arsenals and foreign military factories when rumblings about a coming Civil War reached him at the Red Sea. Though he was getting old for service—Ripley was one of the few serving officers to have been born in the eighteenth century—he hurried home. "Your country needs you," a friend told him when he arrived back in the United States. Ripley responded, "It can have me and every drop of blood in me."

Although he advocated some advanced designs for artillery, Ripley was a staunch foe of repeating rifles, blocking their adoption whenever he could. No need for "newfangled gimcracks" in his view; tried and true single-shooters were all that soldiers required in shoulder-mounted guns. Ripley's resistance to new technology earned him the nickname "Ripley Van Winkle." Spencer had a one-word term for the man: "fossilized." While Spencer can't be blamed for complaining about Ripley's animosity toward repeaters, the Ordnance chief did have his reasons. Gun inventors of all sorts were pitching their brainchildren to the Army in hopes of cashing in with big sales for weapons never tested in battle and which were unlikely to be produced in quantity quickly enough. "A great evil now specially prevalent in regard to arms for the military service," Ripley wrote to Secretary of War Cameron on June 11, 1861, "is the vast variety of the new inventions, each having, of course, its advocates, insisting upon the superiority of his favorite arm over all others and urging its adoption by the Government." Guns such as the Spencer that used special ammunition could also be a problem with supplying troops in the field.

Already, Ripley told Cameron, outside influence had led to the introduction of a variety of weapons no better than the splendid US musket, causing confusion and inefficiency. "This evil," he wrote, "can only be stopped by positively refusing to answer any requisitions for or propositions to sell new and untried arms, and steadily adhering to the rule of uniformity of arms for all troops of the same kind, such as cavalry, artillery, infantry." Besides, he added, there was no one out there who could make as many arms as were needed in less than a year. "Even Mr. Colt, who has the most complete private armory in the United States or probably elsewhere, and greater means and facilities for commencing the fabrication of the Government pattern arms than any one else, states that it will require six months for him to make the first delivery."

Samuel Colt was building on his success with rapidly increasing production, though he had stopped selling guns to the South and declared publicly that he would back the Union. "As he has been for a long time making them [revolvers] for the rebels," *The New York Times* scolded two weeks after the fall of Fort Sumter, "it is time he did something for the United States." Under the headline "A Revolving Patriot," the paper accused Colt of having given Southern buyers a discount he was not offering to the North. "In other words, he sells to the rebels *cheaper* than to the authorities of the Republic. If he could devise any safe way of more effectually aiding the rebellion, Mr. Colt would probably embrace it."

Christopher Spencer and the Cheney brothers were getting attention but had no guns to sell; they didn't even have a way to make them yet. Horace Smith and Daniel Wesson continued to sell as many of their mini-revolvers as they could make and were preparing to produce the larger one suitable for more serious work, perhaps as a backup gun for soldiers. Far behind the others was Oliver Winchester, whose hope for salvation lay in B. Tyler Henry's radically transformed Volcanic. Gearing up to make that gun was still a long way off.

Eliphalet Remington's company was doing well making barrels and other gun parts for soldiers and citizens, but he had stayed away from mechanisms for repeating rifles. Revolvers were another matter. He and his sons dove into the revolver business with enthusiasm, putting out one that Chief of Ordnance Ripley liked enough to order five thousand of them on June 13, 1861, to be delivered "with the greatest possible dispatch." The following month, at the Battle of Bull Run just south of Washington, Confederate forces routed Union soldiers, a sign that the war would not end swiftly. The Remington factory picked up production even more, making physical demands that hit sixty-seven-year-old Eliphalet hard. It may have been appendicitis, though doctors at the time blamed bowel problems, that finally drove him to bed in his home by the Erie Canal in midsummer 1861. There, ever the poet, Lite reportedly turned to verse, dictating to his daughter, Maria, who sat by his side,

In manhood's strong and vigorous prime
I planted a young linden tree
Near to my dwelling, which in time
Has spread its branches wide and free.

Oft have I viewed its healthful growth
With something like a parent's pride
Who sees the offspring of his youth
Grow to strong manhood by his side.

But now, old age had damped the flame
That glowed within me at that day
Energy and strength desert my frame
And I am sinking in decay.

But thanks! I've lived and long have shared
Health and vigor like this tree
And when I'm gone let it be spared
A mute remembrance of me.

On July 12 the founder of E. Remington & Sons died.

The company was now in the hands of the founder's three sons, each bringing useful traits to the job. Philo, the eldest, was a dedicated Methodist, stern and strict like his father, though reportedly ready to lift up the fallen. According to one account Philo approached a Remington worker who kept getting drunk and told him, "Go to church. It will help you to overcome your weakness." But the man was a Catholic; he wouldn't go to a Methodist church. So Philo did what he thought was the next best thing: send the worker a Catholic prayer book, along with a Catholic cardinal's discourse on alcohol indulgence. The eldest Remington brother was more than a moralist. He was good at organization and, like his father, had a gift for mechanics. Samuel, the second Remington son, was a genial bon vivant who enjoyed living well, perfectly suited to be the company's charmingly effective salesman and negotiator, and eager to make money. He had tried his hand at railroad construction in the West, but when that didn't work out, he came back to Ilion, where he opened a store on the banks of the canal in 1845. Samuel made brooms and broom handles, eventually turning his attention to locks and safes, as well as some breech-loading guns, before finally settling down to partner with his two brothers in the Remington business. The youngest, Eliphalet III, was a bookish man of quiet temperament with a fondness for facts and figures. His role was in the company's finances and supervising the office, especially company correspondence, thanks to his fine penmanship and facility with words. The Remingtons' goal was now to do what they could to arm the Union—and compete with Samuel Colt for dominance in the revolver market.

The favorable reports on the Spencer rifle the Cheneys were receiving didn't do them much good with Ripley standing in the way, so in October they hired a lobbyist to push the general. When that didn't work, they dispatched Christopher Spencer himself to Washington on behalf of his rifle. There, on November 4, he showed the repeater to Gen. George B. McClellan, commander of the Army of the Potomac. This led to examination by a trial board, which liked the Spencer. Still no Army contract. It was time, Charles Cheney realized, for more lobbying, which resulted in Assistant Secretary of War Thomas A. Scott urging Ripley in writing to examine the test reports of both the Henry and Spencer rifles and give his opinion on them, "as arms such as these are needed by the government."

Ripley promptly gave his answer: The Henry and Spencer rifles were too heavy; were useless without their special ammunition, which could be damaged when carried on horseback; had spiral springs in their magazines, which might suffer from long use in the field; and were too expensive. And, he wrote directly to Secretary of War Cameron, because "the Government is already pledged on orders and contracts for nearly 73,000 breech-loading rifles and carbines, to the amount of $2,250,000, I do not consider it advisable to entertain either of the propositions for purchasing these arms."

Charles Cheney had too much riding on the rifle to let things drop. Warren Fisher, Jr., the Spencer Repeating Rifle Company treasurer, then wrote to Cameron offering to make ten thousand repeaters at forty dollars apiece to be delivered in lots, the first batch of five hundred in March 1862, followed by installments each month thereafter until the order was completed. That was a deal too good to pass up, despite Ripley's aversion to repeating cartridge rifles. On December 24, 1861, Assistant Secretary of

War Scott put through the order to the Ordnance Department. Ripley was forced to comply.

Two days later the Cheneys finally had a contract to make rifles, lots of them, for the Union, to be used in a war that had been under way since the previous spring. Now they had to find a place to make them—and the machinery and manpower to do the job. They were in the same predicament as Samuel Colt had been when he embarked on making the Walker revolver and as Robbins & Lawrence had been in when they agreed to make thousands of rifles for the government. Three months was not a long time for the Cheneys to make good on their promise. Realistically there was no way they could.

23

CATCHING UP TO WAR: 1862

In January 1862 the Cheneys moved one step closer to production when they incorporated the Spencer Repeating Rifle Company. They were ready to put a chunk of their fortune into making guns, but luring other backers to share the risk was prudent. A formal company was a good way to do that.

Two days after the company came into existence, *Scientific American* published an illustrated article praising "Spencer's famous breech loading rifle, about which so much has been said, and for which large orders have been issued by government for the use of both the army and navy." The article was good publicity, but the Cheneys and Spencer were in no position to provide guns to anyone. First they needed a place to make them. They quickly found it in part of a Boston piano factory.

The Chickering Piano-Forte Company was more than a large enterprise making a particular musical instrument. Chickering was partly responsible for the "piano craze" sweeping the country in the mid-1800s, when a sign of American gentility was a piano in the parlor. "Piano-Fortes are becoming so fashionable a piece of *furniture,*" observed a writer in 1823, "that no house is considered properly furnished, at the present time, unless one of these instruments, polished and gilded in the most extravagant manner occupies a conspicuous place in the principal apartment." The trend was a good one for company founder Jonas Chickering. Not only did he excel at making pianos, he knew how to make money, having proved himself adept at inventing better ways to manufacture the instruments—he patented a method of casting piano frames—and at managing a large business. Like many other successful innovators of his time, Chickering was both craftsman and industrialist, employing an advanced division of labor to a complex manufacturing process. The quality of his pianos was even recognized across the Atlantic, winning him a medal at the 1851 Great Exhibition in London.

Jonas was dead by the time the Cheneys came along, but his sons ran the business as he had. The eldest and senior partner in the firm, Thomas E. Chickering, was also a colonel in the 41st Regiment Massachusetts Volunteer Infantry who was so dedicated to the Union that he offered a two-hundred-dollar bounty for men willing to sign up with his regiment, with an extra one-hundred-dollar bounty for anyone enlisting for three years. These were the kind of people the Cheneys could feel comfortable with.

The year 1862 was still young when the Cheneys converted their rented part of the piano factory to make guns. Christopher Spencer, now living in Boston, designed the machinery required for the work, which made sense, because he was the man who knew the guns best and had invented factory machines before. Time was precious. The Spencer company's March deadline for delivering the first order for repeating rifles was uncomfortably close.

Not everything that comes off an assembly line meets a manufacturer's specifications, of course, and Spencer parts were no exception, including rifle barrels. Trying to shoot a bullet through a defective barrel could be disastrous to the shooter and anyone nearby, but Christopher Spencer managed to find a use for a few of them that were strong enough to funnel steam under pressure. The goal, as usual with controlled steam, was to move a piece of machinery in a desired direction. For Spencer that meant turning shafts that turned wheels on a carriage. A system of ratchets and pawls allowed the outer wheels to turn faster than the inner ones when Spencer rounded corners. With a two-hundred-pound boiler on the back pumping power through rifle-barrel pipes, Spencer's buggy took him around town as easily as if there had been a horse pulling it. And that was the point: no need for a horse, just coal to keep the boiler going. The only drawbacks to driving his automobile on city streets were smoke coming from the coal fire and noise. Clouds from the four-foot smokestack weren't that bad, but the racket made by the two-cylinder engine frightened horses enough for some citizens to complain. Once when a woman couldn't persuade the horse pulling her carriage to go around Spencer's noisy car, he got out, unhitched the animal, led it around the car, and then dragged the woman's carriage to a place where the horse allowed itself to be hitched up again. According to one account, Spencer was responsible for an early automobile accident when he sideswiped a milk wagon.

Perhaps it was a life's worth of hard work mixed with overenjoyment of alcohol that helped sap Samuel Colt's health in the waning weeks of 1861. Rheumatic fever and gout were the leading causes, though, with a Christmas cold contributing. As Colt lay ill, his factory hummed along, making revolvers as well as long arms, including twenty-five thousand rifled muskets he sold to the government the day after Christmas. By the end of the first week of 1862, Colt's mind was faltering, pulling him out of reality and then back in again, while his wife, Elizabeth, tended to him. He told her that he expected to die.

"Then he bade me a last farewell, calling me his faithful, loving wife," she wrote later, "asking me to carry out all his plans so far as I might; and with his last kiss whispered that, when God willed, I should go to him beyond the grave, where partings never enter. Then reason again gave way, and, as he said himself, 'it is all over now.'" Two painful days later, at 9:00 in the morning of January 10, Samuel Colt died at the age of forty-seven of what *The New York Times* called "an acute attack upon the brain."

Colt's funeral was as spectacular as the man himself. More than fifteen hundred factory workers passed two by two through an Armsmear parlor where their employer lay in a metal casket overlain with wreaths of camelias and white roses in evergreens. The Armory Band played a dirge, and a quasi-military ceremonial honor guard called the Putnam Phalanx marched. In early evening Samuel Colt was buried on the grounds of his ducal manor.

At the time of his death, Colt had left his name everywhere: on his company, on the village of Coltsville, and on the flood of guns he sent around the world, including the myriad lushly decorated and encased arms he bestowed on powerful people to induce them into becoming customers. Colt was not alone when it came to purposeful largesse. Oliver Winchester was also in the game, though he never came close to doling out handsome gift guns to the extent that Colt did. However the wood in his tended to be more elaborate.

When Winchester's rifles finally started emerging regularly from the New Haven Arms Company factory in mid-1862, he presented an engraved, gold-plated Henry with a rosewood stock to the man at the top, someone who appreciated new inventions, especially ones that shot bullets. Midway between the gun's rear sight and its operating lever in an oval surrounded by gilt flowers and looping vines on the right side of its mechanism, was this inscription:

LINCOLN PRESIDENT *U. S. A.*

For centuries guns owned by royalty were often richly bedecked with gold, ivory, even jewels, making them markers of wealth and power as well as examples of art worth displaying. While it wore no jewels, Abraham Lincoln's exquisite Henry rifle compared favorably to a king's gun.

Presidential gifts did not have to be reported then, contrary to what today's rules of ethics demand, and it was not generally considered unseemly for a high government official to accept them. Lincoln received many during the campaign and after, from shawls and socks to canes and shirt studs, gold watches, food, a whistle made from a pig's tail, even an expensive suit of clothes to wear at his first inauguration. Sharpshooting soldier and inventor Hiram Berdan fashioned a rifle for the president, a small rendering of Lincoln's head to serve as the gun's hammer. The president did turn down a breeding herd of elephants offered to him by the king of Siam. "Our political jurisdiction," Lincoln explained, "does not reach a latitude so low as to favor the multiplication of the elephant."

Most gifts to Lincoln were simply expressions of respect or support.

The Henry rifle from Winchester, on the other hand, was an attempt to stimulate business for his New Haven Arms Company. There is no evidence that Winchester's present to Lincoln accomplished its mission. Neither, apparently, did the lushly engraved, silver-plated Henry rifle he gave to Gideon Welles nor the Henry he presented to Edwin M. Stanton, who replaced Simon Cameron as secretary of war. Stanton's rifle had the distinction of bearing serial number 1; Lincoln's was number 6.

Even had the fancy gift Henrys seduced their new owners into ordering large batches of their assembly-line cousins, Oliver Winchester's repeating rifles couldn't have been in Union hands in time for the biggest Civil War battle so far, because they were not yet in full production. Nor could Christopher Spencer's repeaters. War, of course, wouldn't wait for them. Neither would the two opposing armies totaling more than a hundred thousand men that approached each other in southwestern Tennessee, as the first week of April neared its end, almost a year after Fort Sumter had fallen into Confederate hands. A one-room log church called Shiloh Meeting House at a small crossroads in Tennessee's wooded tablelands would give the upcoming battle a name.

Over two days, April 6 and 7, 1862, Union and Confederate armies with many green soldiers who had not known combat, clashed over the spring terrain made muddy by rain the days before, and then bloody as men died. "I have heard of wars & read of wars," a member of the Nineteenth Louisiana Infantry wrote two days after the battle, "but never did I think it would be to my lot to participate in such a horrible scene." The Union won, but at a staggering cost. More than twenty-three thousand soldiers had been killed, wounded, or gone missing, a number far greater than US casualties in the entire Mexican-American War.

Riding over the killing ground after the battle, the commander of Union forces, Gen. Ulysses S. Grant, "saw an open field, in our possession on the second day, over which the Confederates had made repeated charges the day before, so covered with dead that it would have been possible to walk across the clearing, in any direction, stepping on dead bodies, without a foot touching the ground." Before Shiloh, Grant and others believed the rebellion would end if the Confederacy lost a significant battle. Now he realized that the soldiers in gray and butternut would not buckle even under torrents of gunfire, choosing instead to fight on in deadly earnest. After Shiloh, Grant recollected years later, "I gave up all idea of saving the Union except by complete conquest."

The Remington brothers were eager to contribute to that conquest. In 1862 they had government contracts to sell a military revolver that they claimed was just as good as Colt's, only less expensive. Colt had been dominant with revolvers when Eliphalet Remington III told an ordnance commission three months after Samuel Colt's death that the family firm could make a Colt gun "quite as cheap or cheaper than our own," but it wouldn't be as good as the sturdier Remington. The Army could have .44-caliber Remington revolvers for twelve dollars apiece, he claimed, if the order was big enough. The implication was that the Colt company had been profiteering. To be competitive with the Remingtons, Colt cut the price of its .44 revolver from $25 to $14.50.

Smith & Wesson was not in the business of courting large government orders—even its new, larger revolver wasn't big enough for military service—but the company gave soldiers the chance to carry an extra gun in case of emergency.

As expected, the Spencer firm couldn't meet its March deadline, but the Cheneys were still getting up to speed, their repeating rifle receiving increasingly favorable treatment in tests.

As for Oliver Winchester, midsummer 1862 saw his Henry rifles go into serious production, though not in the kind of numbers necessary to supply an army, even if the Union Army wanted them, which it didn't. Only about 125 guns came off the New Haven Arms Company's assembly lines in July, followed by the same amount in August. The company squeezed out a few more Henry rifles in September, still not enough to boost its balance sheet into showing a profit. Yet that month would be one to remember, as a September battle would be the bloodiest day of the Civil War, with more soldiers killed on American soil than on any other single day in the country's history. For Oliver Winchester, September 1862 would also be a month for marriage.

24

A BARON'S BID FOR DYNASTY

Winchester made sure to give his son the comfortable upbringing he himself never had. William Winchester's world was not to be hardscrabble farming and poverty topped off by a stepfather's rejection, but one of material amenities befitting an industrialist's heir. Wirt—as the family called him, using his middle name—had two sisters, but he was Oliver and Jane's only surviving male child (another son died before he turned two) and so was destined to take over the Winchester businesses when his turn came.

Until then William prepared for life among the well-to-do, which included graduating from New Haven's boys-only Hopkins Grammar School, one of the country's oldest independent secondary schools. As planned, a bit of travel and time spent working at his father's shirt factory helped ease him into management. The son was not the robust, hard-hitting, hard-charging man the father was. While Oliver's clean-shaven face was strong and stern, William's was wan, his cheeks flanked by whiskers tumbling their way down past his collarbones. His blue eyes were gentle, almost pleading, his auburn hair soft. It was said he had "a retiring disposition."

He also had prospects, so it was natural that twenty-five-year-old William Winchester and shy, petite twenty-three-year-old Sarah Lockwood Pardee found each other. The Winchester and Pardee families were neighbors, and both patriarchs were officers of the First Baptist Church of New Haven. Like Oliver Winchester, Sarah's father, Leonard, had patented something, a means of joining wood staves for construction, six months after Oliver patented his shirtmaking system. Now Leonard Pardee was a successful maker of "all kinds of architectural carving," including sashes, doors, blinds, and piano legs, and he had for a few years in the 1840s run the city bathing house. The Pardees were well off—not Winchester well off, but respectably set. Sarah's life had been filled with the benefits of modest wealth and proper breeding. Education befitting a daughter of means, including music instruction (she was deft at the piano) and tutoring in French, had given her a due measure of refinement, her considerable intelligence making her all the more interesting—and desirable. It was no surprise that William had fallen in love with her.

The wedding was set for Tuesday, September 30, 1862, with summer just past and New England trees showing early fall colors. A social event of such importance, though not lavish, made the Civil War seem distant, even though the bloodiest day in American history had ended only thirteen days before, in the fields and woods of western Maryland flanking Antietam Creek. More than twenty thousand men had been killed or wounded in fighting that started at daybreak and continued until nightfall. A farm lane that ran through part of the battlefield, worn down by years of weather and wagon traffic, was called the Sunken Road before the battle. After thousands of soldiers died there, it had a new name. It was now Bloody Lane.

Even though Robert E. Lee's Army of Northern Virginia was not destroyed at Antietam and managed to slip away across the Potomac River, the battle was enough of a victory to give Abraham Lincoln the occasion to issue his preliminary Emancipation Proclamation. Lincoln had wanted to issue it earlier but was urged not to in case it might be viewed as a desperate attempt by the Union to stave off defeat. Wait for a major battlefield victory, some members of his cabinet counseled. Antietam gave the president that opportunity, which he took five days later by declaring that slaves held in states still in rebellion on January 1, 1863, would be "then, thenceforward, and forever free." Freedom for slaves, in addition to saving the Union, became a reason to keep fighting.

"Everything now is quiet," wrote George Washington Whitman, poet Walt Whitman's soldier brother, to their mother on Sarah Pardee's wedding day, "and it is quite a releif [*sic*] to be out of the sound of canon [*sic*] after hearing it almost daily, and sometimes nightly, for two or three weeks." As he put his thoughts onto paper, George Whitman's temporary home was a light canvas, two-man tent in the mountains several miles from the stilled battlefield where he had fought and thousands had died. A photographer and his assistant from Mathew Brady's studio had come to Antietam two days after the battle and recorded images of men lying where they had died, giving the public a first look at what the aftermath of real conflict looked like before the dead are buried. Their photographs, in the stark monochrome of the time, showed none of the heroic glory so often seen in paintings.

Despite the calm, dying from the Battle of Antietam continued on September 30, as men succumbed to infection in wounds that couldn't heal in bodies too weak to resist. Elsewhere in the crippled nation that day, more gunfire claimed more lives, including the nearly four hundred men who fell fighting in southwestern Missouri in what came to be called the first Battle of Newtonia. It was also a Civil War battle in which Native Americans fought on opposing sides.

The tide may have turned in the Union's favor at Antietam, but no one thought the war was about to end. There would be a need for more and better guns. Had Union infantrymen shouldered Winchester's Henry rifle, would the Battle of Antietam have ended more quickly? Maybe Robert E. Lee and his surviving Confederate troops would not have been able to flee so easily across the Potomac River into Virginia. If his infantry had carried Henrys, maybe overly cautious Union general George B. McClellan would have been emboldened to follow Lee, decimating the Army of Northern Virginia with rapid fire and shortening what looked to be a long war. This was raw speculation, though perhaps a potential selling point for Oliver Winchester. There would have to have been a lot of Henry rifles to speed the war to an end, though, far more than he had the capacity to produce.

But the matter at hand was getting Oliver's son married. For the moment guns could wait. Until their own home could be built, the new Mr. and Mrs. Winchester moved into the patriarch's palatial estate on appropriately named Prospect Street, not far from the shirt factory. They would stay there for several years.

With Sarah now part of the Winchester family, and the expectation that his son's marriage would produce an heir, Oliver Winchester prepared to travel again to Washington to peddle his wares. More battles lay ahead. When October began, the New Haven Arms Company was on its way to turning a profit, its assets approaching the value of its debts. One immediate but delicate challenge for Winchester was reining in a valuable but troublesome agent out west whose behavior threatened to endanger the company's future.

25

THE HENRY OR THE SPENCER?

By early October 1862 Oliver Winchester had had just about enough of George Dennison Prentice.

Winchester's network of dealers—which included Kentucky's George Prentice—usually did what they were told when it came to pitching Henry rifles to customers: Make honest representations about the marvelous new repeater, show how it was better than yesterday's single-shot gun, and sell each at the price specified by Winchester's New Haven Arms Company. But when Confederates were poised to invade his town that September, Prentice panicked and sold part of his allotment to Union loyalists below cost, much to the irritation of other agents in nearby states, who quickly accused Winchester of having too cozy a relationship with his man in Louisville.

"We will not sell or furnish any parties with the rifles who does not adhere to the prices," Winchester scolded Prentice in an October 8 letter. "They are low enough to be just to all."

With the war heating up and his company struggling to find solid financial footing, Winchester didn't need headaches from people like Prentice. But he still needed Prentice, who was not only an agent for the Henry rifle but also the powerful founder and editor of the widely read *Louisville Daily Journal.* Prentice had praised the Henry in print as "the most beautiful and efficient rifle we ever saw." The newspaperman made no secret of working on Winchester's behalf. He told his readers so with notices printed in his paper starting on August 30, 1862: "The agency for this weapon has been taken from Mr. W. C. Stanton by the Senior Editor of the Journal. To him good Union men may apply. If others apply, he will lodge information against them as suspicious characters. Let all true patriots send in their orders." The Henry rifle was selling briskly, Prentice proclaimed, because every "man in possession of it is a garrison, every company an army. In these terrible times no loyal man should be without it."

The Journal kept the sales pitch going. On October 2, readers were informed:

> The continual calls for the Henry rifle have carried off nearly the whole lot on hand. We know not when more can be had, as the demand is very great all over the country and the supply at the factory is very limited. Let every man call immediately, who, in these times of peril, would make himself equal to fifteen men, nay thirty or forty. Call immediately.

Prentice's enthusiasm was welcome; his below-cost selling was not. But Winchester respected the old adage: Don't pick a fight with someone who buys ink by the barrel, especially a man with a savage wit skilled at spinning editorials that bit. Besides, Prentice was no ordinary newspaperman. His pungent, often sarcastic style brought to *The Journal* readers far beyond Kentucky's borders. Navy Secretary Welles referred to him as "the distinguished and patriotic journalist." Thanks to Prentice's guidance, *The Journal* had become the most widely circulated newspaper beyond the Appalachian Mountains. "Mothers named their children for him," noted one historian, "poets and poetesses honored him in verse." Paragraphs that flowed from the Prentice pen, and later from dictation necessitated by his crippled, overworked hand, found homes in newspapers in both the United States and Europe under a column titled "Prenticeana." Salivating at Prentice's power and reach, Winchester would nevertheless continue to court him by sending him an advertisement for the Henry rifle to be inserted in both daily and weekly editions with the request, "Please display & make the same as attractive as possible. An editorial notice, calling attention to it, will be duly appreciated. . . ." He also gave Prentice a Henry rifle.

Perhaps it also made sense for Winchester to cut the editor a little slack after what he had just gone through. Despite Prentice's passionate opposition to secession in a state filled with divided loyalties even within families, his two sons had bucked their parents and joined the Confederate Army. Adding to Prentice's torment, his elder son, twenty-five-year-old William Courtland Prentice, was killed by friendly fire one week before Winchester chastised him in the October 8 letter.

"This young man, if he had always directed his energies judiciously, could have made himself a distinguished ornament in any profession of life," wrote Prentice, pouring out his anguish in *The Journal* after learning of his son's death. "But an intense Southern sympathy, in spite of the arguments, the remonstrances, and the entreaties of those who dearly loved him, made him an active rebel against his country." In a letter to a friend a few weeks after Courtland was killed, Prentice wrote, "I am in bad spirits, my dear friend, for my own sake and our country's. My son is dead, and sometimes I almost fear that my country, too, may perish. I see no palm-tree upon the desert which surrounds me."

Also, widespread fear that rebel forces were at the gates of Louisville was justified. A Confederate Army was indeed preparing to attack the city, leading Union general William "Bull" Nelson to issue an evacuation order the day after Prentice's son fell in battle. In that order Nelson spoke with urgency: "The women and children of this city will prepare to leave the city without delay." All residents who choose not to fight, Nelson warned, must stay in their homes or risk being shot.

Prentice was no coward. In 1857 he had shot it out with the editor of a rival paper in a pistol duel. (The only casualty was reportedly a bystander hit in the leg by Prentice's bullet.) But in the fall of 1862, his entire city faced danger. Breastworks and entrenchments were being built around Louisville to wall it off from siege. Residents were fleeing. Better to make sure that the remaining citizens loyal to the Union could arm themselves with rapid-fire Henrys, Prentice thought, even if that meant breaking his agreement with Oliver Winchester. And he didn't want those fast-firing rifles to fall into enemy hands. Traditionally tough and demanding with his agents and pretty much everyone else, Winchester still continued supplying Prentice with rifles and parts, even as Prentice kept selling below cost.

Winchester wrote him again on October 25:

> We have to beg that you will not sell any more rifles at the prices you sold the last that you had. It has so discouraged some of our best customers in the neighboring states, that they have declined buying them. We perfectly understand that you pursued the course you did, in selling them without a profit, and even at a loss, from purely disinterested motives; but the result has proved very injurious to us, as regular dealers will not be convinced that we did not give you some decided and unfair advantage over them.

Winchester's New Haven Arms Company secretary tried to mollify a Henry rifle seller in Indiana by writing, "We regret the mischief done by Prentice and others in Louisville. We were not aware of what they were doing until the mischief was done." Their only excuse, the company secretary went on, was they wanted to enable

> any man who wanted one to defend himself during the expected attack on Louisville to supply himself. There has not *one been on sale there* for nearly a month; nor shall we sell any more there until we are satisfied that they are going into hands that will adhere to the prices. We shall endeavor to maintain, to the best of our ability, the uniformity of prices as a matter of interest to ourselves, and duty to all fair dealers.

Winchester personally complained to two Kentucky dealers in mid-October that "Mr. Prentice, we have no doubt, with a most disinterested purpose, sold the rifles we sent to him for less than cost. . . ." This was bad business that "will be very injurious to us in the end."

The attack on Louisville never came, and Oliver Winchester made several trips to Kentucky to push his agents and inspire Union troops to buy Henry rifles. Although individual units were acquiring Winchester's gun, as were civilians, and the Army finally issued limited orders, big contracts continued to elude him. New Haven Arms Company investors were getting nervous. They wanted some return for the money they put into the business and worried about how much, if anything, they would get. Winchester hoped they would fork over more cash to keep the firm running, warning that they would lose money if the New Haven Arms Company went under. He even floated the idea of pulling out of the business himself.

"The assets of the Company cannot be sold for enough to pay its indebtedness," Winchester wrote to E. B. Martin, one of the shareholders, on October 17, 1862. "I have repeatedly offered to give all of my stock, to the other stockholders, if they would reimburse my advances, and relieve me from my responsibility for the Company, which covers its whole indebtedness." That indebtedness, he told Martin, totaled $77,437, including cash advanced by Winchester personally. "So much for the *present* value of the stock, which you will see is nothing," he continued. "The stock, however, has a value, but it is entirely prospective. It arises from the fact that our new rifle is a success, and will, in time, if pressed with vigor, retrieve our past losses; but to do this, further aid and support will be needed from the stockholders."

If Martin decided to go ahead and sell, Winchester warned, he would get only twenty-five cents on the dollar for his stock. "We should prefer to have you hold on, and help us."

Problems also plagued the Cheneys. No Spencer repeating rifles would be ready for distribution to the military for several months, although a number of specially made rifles had been used in Army and Navy trials. What the company did have was an effervescent salesman in the inventor himself. During the first weeks of 1863, Spencer made plans to visit troops in the field as a traveling salesman for his rifle, getting to his destinations aboard trains and on horseback. In February he headed for the Army of the Cumberland, then encamped near Murfreesboro, Tennessee, a state embroiled in conflict.

With him was a sample Spencer rifle and a supply of ammunition, plus an account book to record his travel expenses for the company. While Samuel Colt had a talent for spending with abandon, Spencer was a tightwad:

- No allocation for breakfast
- 50 cents for lunch
- 25 cents for supper
- One dollar per night for sleeping-car accommodations when traveling by train
- For entertainment and tips: Nothing

Once with the troops, Spencer showed as many potential customers as possible—corps commanders, mostly—how to handle his rifle and how to load it, all the while extolling its man-killing virtues with missionary zeal. Gen. William S. Rosecrans, known to the soldiers as "Old Rosey," had favored Colt's revolving rifle but took a liking to the Spencer and placed an order for several thousand with the Ordnance Department back in Washington. That meant dealing with bureaucracy and stubborn General Ripley.

Oliver Winchester was also trying to make inroads at the front, but through agents touting the excellence of the Henry rifle. He already had one convert in Murfreesboro: a thirty-three-year-old colonel named John T. Wilder, a foundry owner and inventor of hydraulic machines. Wilder's mechanical savvy led him to consider breech-loading repeaters as the ideal weapon for his command. So enthralled was he with the Henry that in March he ignored proper channels and wrote directly to the New Haven Arms Company:

> Gentlemen:
> At what price will you furnish me nine hundred of your "Henry's Rifles," delivered at Cincinnati, Ohio [where the Union Army had a large depot], without ammunition, with gun slings attached? Two of my regiments, now mounted, have signified their will-

> ingness to purchase these arms, at their own expense, if they do not cost more than has been represented to them. My two other regiments will be mounted soon, and will, doubtless, go into the same measures. It is of course desirable to get them at as low figures as possible, as the men are receiving from the government only thirteen dollars per month. How much additional expense would it be to have an extra spiral spring for each gun, to replace any that may be broken, or when worn out?

Unfortunately for Winchester, his company couldn't supply that many Henrys all at once, and Wilder was in a hurry.

Taking advantage of Winchester's inability to supply Wilder with Henry rifles, Christopher Spencer approached the colonel. The two men were an odd-looking pair: Wilder, a colossal figure in his blue uniform, six foot two and weighing two hundred ten pounds, dwarfed the slight five-foot-five Yankee in civilian attire. But both were passionate inventors who appreciated how bits of physical objects could be made to do more than they had before, and they were eager for a cause they cared about. Wilder had gone into gunmaking himself, though not of weapons a soldier could hold. Just before the war broke out, he had designed and cast a pair of wrought-iron cannons at his Indiana foundry. Wilder and Spencer surely saw in each other a man of ideas, someone to be respected.

It didn't take much for Wilder to be sold on both the inventor and his novel rifle. Things were likely to move quickly in the field, so the colonel took the same action with Spencer as he had with Winchester. Rather than wait for the military bureaucracy to process orders, Wilder offered to buy them directly from the Spencer company with loans from his hometown bank in Indiana. He suggested that his troopers pay for them, thirty-five dollars apiece, a sum that was nearly three times the monthly salary of an enlisted man. Wilder would cosign bank loans for each one. Believing in their charismatic commander and the new rifle, the soldiers agreed, though they were spared the need to dig into their own pockets when the Ordnance

Department sent them Spencer rifles anyway. With much excitement and some anxiety, soldiers waited for the new guns to arrive.

By mid-May their Spencers showed up, an event local newspapers thought worth announcing to the world. "Col. Wilder has just received the long-expected 'Spencer' rifles, the most excellent gun for mounted men yet manufactured," trumpeted *The Evansville Daily Journal.*

"We have drawn new guns and they are the nicest and handiest gun I ever saw," a Union corporal wrote in a letter to his family. "They are called the spencer repeating rifle. They shoot seven times and can be loaded and fired in a little less than no time." As soon as the men were handed their repeaters with allotments of cartridges, they dashed into the woods, trying out the "seven shooters" on rabbits, squirrels, turkeys, and pretty much any small game they saw. They were delighted by the rugged rifles that could shoot repeatedly. Single-shot muzzle-loaders, one Indiana soldier concluded, were now too slow "for a *fast* people like the Americans. . . ."

Wilder's scouts chose not to use Spencers; instead, they bought Henry rifles, because they could be fired with greater speed and held more than twice as many rounds as the Spencer. They were perfect for shooting one's way out of a tight situation often encountered by scouts, so they thought.

While Spencer was in the field pushing his rifle, Oliver Winchester was pushing his own in print, taking General Ripley to task by name. The advantages of rapid-fire weaponry were so obvious, Winchester wrote in *Scientific American,* identifying himself only as "O.F.W.," that there should be no question about them. "Such, however, is not the case. General Ripley, Chief of the Ordnance Department at Washington, opposes their introduction into the army, and has recently said that he prefers the old flint-lock musket to any of the modern improved fire-arms, and that he believes nine-tenths of the army officers will agree with him." Winchester argued that repeaters would give confidence to the men who used them for combat and terrify the men they shot at.

"The only reason, or excuse rather, we have ever heard against the use in the army of arms susceptible of such rapidity of loading is that the troops would waste the ammunition," he wrote, taking aim at another of Ripley's objections. That was poor logic, Winchester believed.

> If, *to save ammunition*, it is essential that every soldier should remain for sixty seconds while re-loading, a helpless target, to receive his opponent's fire of from one to fifteen shots, why not reverse the order of progress and turn the ingenuity of inventors to the production of a gun that will require twice the length of time or more to re-load, and thus *double the saving of ammunition?* Saving of life does not appear an element worthy of consideration in this connection. Yet this is West Point opinion—the deductions of West Point science! Are these results worth their cost to the country?

In early April 1863 Winchester considered taking a different approach to challenging his competition. He had been doing business with William Read & Sons, a Boston gun and sword company less than four hundred yards from where Spencers were being made. "Will you obtain a Spencer rifle for us," he asked Read in a letter. "We wish to submit to [counsel] for consideration of the point which we think infringes upon ours. I presume we could not get one from them direct." The Spencer rifle was patented before the Henry and for the most part functioned differently, but perhaps Winchester thought some part of the Spencer might be an infringement of a right he had acquired before the Henry was conceived. Would he have wanted to stop production of the Spencer in the middle of a war, or would he have wanted a share of the profits? Probably the latter, because what loyal Union man would want to hurt the war effort? Whatever reason Oliver Winchester had for weighing whether to wage a patent fight over the Spencer, he didn't follow through.

On Wednesday, June 24, Winchester sat at his desk in New Haven dictating two letters to Washingtonians who mattered. One was to Col. Lafayette C. Baker, in charge of the First District of Columbia Cavalry. It was a report giving the status of an order for Henry rifles that month, a mundane one except for telling Baker, "Our Cartridge works blew up some weeks ago in consequence we are out but will ship you four thousand [cartridges] per day until your order is filled. I shall be in Baltimore on Friday and may call on you in Washington."

The other letter sent to Washington, filled with passionate advocacy, was addressed to someone higher than Baker: Winchester's and Spencer's perennial nemesis, General Ripley. Winchester didn't mention to Ripley his problems with Prentice in Kentucky or with other New Haven Arms Company agents who didn't toe the company line. Nor did he say anything about his failure to provide Wilder with the Henry rifles the colonel had so desperately wanted, thus losing a business deal to Spencer. Instead he extolled the Henry's magnificence in an effort to ease the general's resistance to buying the repeaters for combat and perhaps stimulate enough interest to finance expansion of his company. Winchester was taking advantage of his small victory in supplying the DC Cavalry with Henrys, and he wanted Ripley to notice. "[I]f these arms are used as efficiently by the men who are to receive them as they have been by our Union friends in Kentucky," Winchester wrote, "the country will have no cause to regret the expenditure."

An hour and a half before dawn on the day Winchester dictated his letters to Baker and Ripley, a bugle sounded reveille, waking John T. Wilder's fifteen-hundred-man Hatchet Brigade. They were called that, because of their habit of carrying small axes with two-foot handles as multipurpose tools and close-combat weapons. The men quickly got up, ate breakfast, and readied themselves to move as a unit before daylight.

Wilder had an air about him that announced he could lick whatever foe he faced; the soldiers under his command tended to agree. Wilder's troopers, mostly farm boys plus a fair number of laborers, held together by their charismatic commander, carried their new Spencer rifles, along with eighty rounds of ammunition each. The guns had arrived only a few weeks before, but the men had rapidly learned to respect their newfound firepower.

A dozen miles away was Hoover's Gap, a narrow valley that wound through a line of wooded hills three hundred feet high and thick with underbrush and briars. General Rosecrans wanted to bring Hoover's Gap under Union control. Down the valley ran the Manchester Pike, a road barely wide enough for two wagons, but it was macadamized, suitable for travel wet or dry, unlike the area's dirt-covered routes. Hoover's Gap was one of several passages through the ridges and hills that made up the fertile valleys of Middle Tennessee, whose strategic importance to the Confederacy was crucial. By controlling that region, the approaches to Chattanooga—a key transportation center for the Confederacy—were protected. If the Confederacy lost Middle Tennessee, the Union could move on Chattanooga and then into Alabama and Georgia. The area also provided vital forage and supplies for Southern armies. Late June was a good time for soldiers to supplement army rations, as the hills around Hoover's Gap were rich with ripe blackberries and huckleberries. Apples were abundant, and peaches were nearly ready to be picked. The weather had been dry and gentle, giving the region a soft glow of early-summer calm. Soon that would change.

Union strategists decided that control of Hoover's Gap was critical. One general wanted a cavalry brigade sent along to support Wilder's advance. If the colonel and his men went too fast into Hoover's Gap, the general worried, they might get captured by the larger Confederate force nearby. If that happened, hundreds of Spencer rifles with ammunition would be in enemy hands, ready to unleash their firepower on Union soldiers. Not to worry, the general was told; Wilder can handle what may come.

As the Hatchet Brigade rode from camp, the long dry spell ended with a drizzle. The morning was cool and dark, the quarter moon having set shortly before midnight. Rain-heavy clouds barred light from the coming sun. As the mounted troopers passed through other Union encampments, they heard the sounds of an army stirring. Drums rattled, bugles sounded, and wagons clattered, telling them that the entire Union force would soon follow in their wake. "[W]e knew full well, from the direction we were taking, that a few hours march would bring the brigade to some of the strongholds of the enemy," a trooper later remembered, "so there was silence in the column as we moved along through the mud, and every ear was strained to catch the sound of the first gun of our advance guard that would tell us of the presence of the enemy."

Dawn was slow to arrive, the drizzle having become a steady rain as the brigade moved forward. One trooper griped sarcastically:

> The boys said it was all "Old Rosey" was waiting for, as he did not like to march us on such dusty roads for fear of spoiling our complexion. . . . [O]ur armies seldom if ever advance when the roads are dry and passible [sic], but as soon as the mud is knee deep, as soon as artillery and baggage wagons will stick fast in every mud hole, then the order is given to advance rapidly.

It was "no Presbyterian rain," another soldier said, "but a genuine Baptist downpour." Still, the men were mostly cheerful, at least more cheerful than the weather.

The men heard the first gunshots around noon. The brigade picked up its pace despite torrents of rain pounding them, because Wilder wanted to surprise the enemy by punching into Hoover's Gap and holding it until the rest of the Union Army arrived. On their way to the Gap, the Hatchet Brigade charged through a Confederate cavalry regiment that

> was so much surprised by our sudden appearance that they scattered through the woods and over the hills in every direction, every fellow for himself, and all making the best time they could bareback, on foot and every other way, leaving all their tents, wagons, baggage, commissary stores and indeed everything in our hands, but we didn't stop for anything, on we pushed, our boys, with their Spencer rifles, keeping up a continual popping in front. Soon we reached the celebrated "Gap" on the run.

The Confederate retreat was so sudden that the regiment's embroidered silk banner was abandoned on ground turned into a quagmire by the drenching rain.

Wilder's men swept through the valley at full gallop until they spotted a white puff of smoke from a hill a half mile in front of them. Then came a dull, heavy roar followed by the shrieking of a Confederate cannon shell. The troopers had gone as far as sane men would and six miles farther than they had been ordered to advance. Captured rebel soldiers told Wilder that he faced four infantry brigades and four batteries of artillery. The bulk of the Union Army lumbered twelve miles behind them, its progress slowed by ankle-deep mud. Could Wilder hold out against a vastly superior force? Maybe the general who had worried that the Hatchet Brigade might be captured was right after all. Wilder kept the brigade in place, settling his troopers in lines, where they waited for the Confederates to make a move other than lobbing artillery rounds in their direction, shells screaming "so close to us as to make it seem that the next would tear us to pieces." In time Confederate cannon fire subsided, but not before a shell decapitated the Seventy-second Indiana Infantry's thirty-three-year-old chaplain, who had joined the regiment one week earlier and preached his first sermon the day before the battle.

The Confederate Army of Tennessee prepared itself for the attack it knew it had to make if it was ever going to wrest Hoover's Gap from the enemy. At least one command could be counted on for bravery under fire: the Twentieth Tennessee Regiment Volunteer Infantry. And to advertise it they carried a battle flag like no other. Its banner was made of white and red silk embroidered by Mary Breckinridge, wife of Confederate general and former US vice president and 1860 presidential candidate John C. Breckinridge, from dresses she wore at her wedding and on the day after. She had insisted that the flag, with a carved eagle atop its staff, be given to the most gallant regiment in the division, a responsibility as well as an honor, because enemy soldiers liked to shoot standard-bearers and capture their flags. The Twentieth Tennessee had served valiantly under Breckinridge at Shiloh. The general himself had been with his men in Tennessee until a month before Wilder's brigade swept into Hoover's Gap.

The men of the Twentieth Tennessee were the natural choice to be awarded the Breckenridge banner, as the regimental commander had recognized when presenting it to them three months earlier. "Soldiers," he proclaimed, "to you I commit the gift; in its folds rest your honor. Let it never be contaminated by a foeman's hand."

It was midafternoon when an advance unit of Confederates launched its attack. One of Wilder's units, the Seventeenth Indiana Volunteer Infantry Regiment, spotted the butternut-colored uniforms of the Twentieth Tennessee through rain and smoke from artillery fire that had just ceased. The Tennesseans were moving toward them up a gentle slope through thick undergrowth, but the Hoosiers held their fire until the rebels charged from a hundred yards out, shooting and trying to reload their smooth-bore muzzleloaders as they advanced.

"'Steady.' The word came from lip to lip along the line," remembered a Union captain, "while the yell of the enemy grew nearer and nearer, and the men awaited breathlessly the quickly following order, 'Fire!'"

Spencers erupted from the line of kneeling soldiers, sending a torrent of lead into the advancing Confederates.

Almost immediately after the shooting began, a Confederate bullet slammed into the chest of Christopher Columbus McReynolds, a twenty-year-old corporal in the Seventeenth Indiana, who dropped to the ground, his mission no longer to shoot but to disable his Spencer, rendering it useless if it ever fell into enemy hands. All Wilder's men understood the danger if their rapid-fire rifles were turned against them. McReynolds lacked the strength to do the damage he wanted, so he took out his knife, unscrewed part of the rifle's mechanism and threw it as far as he could. Knowing that his Spencer was now useless, he fell back and died while bullets whizzed about him.

The men of the Twentieth Tennessee reeled from the Seventeenth Indiana's first Spencer volley, which shattered the Breckinridge banner's staff and blew the eagle off its top, sending the silk flag to the ground, where it was quickly recovered and held aloft as the Tennesseans resumed charging the men in blue, giving full-throated rebel yells as they ran. They hoped there was enough time to reach the Union line before the Federals could reload. They were wrong. They hadn't counted on facing Spencer repeating rifles.

Working the levers of their Spencers repeatedly as though they were pump handles, the Union men sent a second volley ripping through the charging Confederate line, then another, and another. The rebels retreated, some crawling to safety, leaving dead and wounded on the field, not understanding why their comrades had been slaughtered in such numbers. Afraid but not in a panic, the men in butternut swiftly pulled back with still more bullets from Spencer rifles hurtling after them.

"[B]ut few men of that 20th Tennessee that attempted the charge will ever charge again," a Union officer said in a letter to his wife after the battle.

"The effect of our terrible fire was overwhelming to our opponents, who bravely tried to withstand its effects," Wilder wrote. "No human being could successfully face such an avalanche of destruction as our continuous fire swept through their lines."

The Spencers' all-but-inexhaustible firepower led the Confederate brigadier general in charge to believe that he faced a force far larger than it actually was.

By late afternoon the rest of the Union Army was close enough for Capt. A. A. Rice, Gen. Joseph F. Reynolds's adjutant, to ride up with orders calling for the Hatchet Brigade to fall back. Wilder refused, saying his men would stay where they were, "as there was no danger of our being driven out of the position." Reynolds should come up and see the situation for himself, Wilder said; if he were here, he would not order me back. With that, Rice threatened to arrest Wilder, but the colonel held firm.

Eventually Old Rosey himself, Reynolds's superior, rode up with some of his staff and asked Wilder what he had done. "General Rosecrans took off his hat and handed it to an orderly," Wilder recalled, "and grasped my hand in both of his, saying: 'You took the responsibility to disobey the order, did you? Thank God for your decision. It would have cost us two thousand lives to have taken this position if you had given it up.'"

"This engagement thoroughly tested the power of the Spencer rifles and proved their great superiority to the muzzle-loader," Wilder wrote after the battle. "For us it did more, it inspired us with a confidence in ourselves which of itself was worth double our numbers. Ever after the brigade would cheerfully have fought ten times its own strength."

The Hatchet Brigade's dash to and through Hoover's Gap meant that it further deserved another name it had been earning. It became known as Wilder's Lightning Brigade.

The Battle of Hoover's Gap was the first time an entire brigade entered combat armed with repeating cartridge rifles, setting a new standard for bloodshed. Rapid fire turned back a superior force, cutting men down with an unheard-of level of efficiency and opening the way for Union forces to reach Chattanooga. Mass killing had taken another step forward.

The week after the Battle of Hoover's Gap, Christopher Spencer's destructive marvel would get another chance to prove itself near a Pennsylvania town called Gettysburg, this time in the hands of volunteers led by a flamboyant young cavalier soon to be nicknamed The Boy General of the Golden Locks.

26

GETTYSBURG

"To you I should have written long ago," Maj. Noah Ferry told his aunt Mary, the first schoolteacher in his Michigan hometown and the woman who had taught him when he was a child, "but having been upon the drive for the last two weeks I have really had no opportunity so to do, at the length I wished; and shall now have to scribble with a pencil." Sitting beneath an apple tree in an orchard near Hanover, Pennsylvania, Ferry continued:

> Yesterday [June 30, 1863] the 5th Michigan [Volunteer Cavalry regiment] had their first smell of battle near Littletown and behaved finely. Our loss was but one killed, while fifteen dead rebels lay in front of our line upon a single field. Had it not been for an order about 4 o'clock P. M., to retire to the position occupied in the morning, I think we should have captured between one and two hundred rebels, whom we had nearly surrounded. Our part of the affair was decidedly the most brilliant of the day.

On that day the Fifth Michigan saw its first serious action of the war, something its men had wanted for months. They had been in Washington days earlier, saddled with more mundane duties. There they had trained, served as bodyguards for visiting officials, scouted, performed picket duty, and occasionally tangled with guerrillas. At least they had first-class weaponry then, thanks to hard lobbying by Michigan governor Austin Blair and other home-state officials.

A deep-in-his-core abolitionist, zealous reformer, and devoted Union man, Governor Blair had long loathed the breakaway South's slavocracy. For years he had pushed the Michigan legislature to give African Americans the right to vote, but that was a concept too radical for the times, so legislation he had backed as a state senator in 1855 never reached the senate floor. At the beginning of 1862, with war under way and Blair having been governor for a year, the state's legislators supported him by resolving "that as between the institution of slavery and the maintenance of the Federal Government . . . slavery should be swept from the land, and our country maintained." Lincoln was then still eight months away from issuing his preliminary Emancipation Proclamation.

Impatient for the Lincoln administration to take more serious offensive action, Blair wanted his Michigan soldiers to have the best guns when that action came. He had regularly thrown himself into providing what he thought the soldiers needed—visiting troops in hospitals and on the front lines, including Shiloh shortly after the battle—and found that he and other Michigan officials had to battle Washington if they were going to put Spencer repeating rifles, the latest thing in firepower, into the hands of their state's soldiers. Once again General Ripley had been a bureaucratic barrier that Blair and his allies had to break through. Lobbying persistently, they succeeded.

Thanks to help from Secretary of War Edwin M. Stanton, an initial order of five hundred Spencer rifles was completed and ready to be sent by speedy passenger train from the Cheneys' Boston factory to Detroit on November 22, 1862. Only the rifles didn't start their journey west until early December, and then it took them three days to arrive. When the Spencers finally did show up, they were one day too late; by then the Fifth Michigan Cavalry was on its way to Washington. After a quick turnaround, the new repeaters made the trip back east and finally found themselves with the Fifth Michigan. "At last we are soldiers, well drilled and well armed," wrote one cavalryman in a letter to his wife on January 6, "because we have received our rifles yesterday, and what is even nicer is that they are breechloaders [and take] 7 cartridges at a time . . . and a saber on top of that. Finally, there is not a regiment in Washington as well armed as we are."

The Fifth Michigan Volunteer Cavalry regiment was a proud outfit. As the name confirmed, its members were volunteers, and not a bunch of ragtag refugees fleeing to the military as an escape from poverty or unsavory lives. As well as farmers and mechanics, among its ranks were businessmen and lawyers, men who had left lucrative careers to become soldiers in the righteous crusade of restoring the Union. Noah Ferry fit that profile. He had stepped away from a prosperous lumber business he ran with his brothers in order to put on a uniform and burnish the already great respect he had earned back home. Despite his devotion to the Union, the decision was not an easy one for the thirty-one-year-old Ferry, because he felt a strong duty to his family as well as to his country. Country won out. Like the rest of the Fifth Michigan, Ferry had tired of the humdrum of military life around Washington. They wanted to fight. Their chance for that came when Robert E. Lee's Army of Northern Virginia invaded the north for a second time in the early summer of 1863.

On June 25 the Michiganders left the capital for the kind of war they had signed up for. Five days later they were up against not a bunch of guerrillas, but thousands of regular cavalrymen led by a legendary master of mounted troops, Confederate Major General J.E.B. Stuart.

The men of the Michigan Brigade, which included the Fifth Michigan Volunteer Cavalry regiment, followed their own legend-in-the making, a brash twenty-three-year-old brigadier general who only a few days earlier had been a mere captain. George Armstrong Custer stood out from other senior Union officers not just because he was young, fair-complexioned, and had ringlets of golden hair, perfumed with cinnamon oil, cascading toward his collar, but because of his custom uniform. He wore a brocaded black velvet jacket with gold piping on the sleeves, a scarlet neckerchief, and a soft wide-brimmed black hat "adorned with a gilt cord, and rosette encircling a silver star," its brim "turned down on one side giving him a rakish air." His spurs were gilt. The newly minted general was "one of the funniest-looking beings you ever saw," wrote a staff officer, "and looks like a circus rider gone mad!" Custer was a man to be noticed, which is what he wanted. The Wolverines—

the Fifth and all the Michigan cavalry regiments—had noticed also that their boy general could fight. He would be out front, in the midst of the action, often ahead of it. This was a leader they could follow.

Noah Ferry's moment of ease in the apple orchard was brief. Before he could tell Aunt Mary everything he wanted to say, word came that meant letter-writing time was up. "We are ordered off again," he wrote, "and I must close without finishing."

About fourteen miles west and slightly north of the orchard lay Gettysburg, a Pennsylvania town where ten roads converged from all points of the compass. Home to twenty-four hundred citizens, Gettysburg had an assortment of businesses, including carriage makers and tanneries. It was there that two massive armies had begun battling each other and where Ferry and the Fifth Michigan were headed on a hot, dusty July 1. As the Wolverines started on their way, the boom of distant cannon fire reached their ears.

The Fifth Michigan would hear the fighting on July 1 but wouldn't be part of it. The same would be true for July 2. Between three and four o'clock in the morning of Friday, July 3, the troopers arrived at a crossroads southeast of Gettysburg, where they wearily dismounted. "Come let us lie down and get a little sleep," Ferry said to his friend and fellow officer, Maj. Luther S. Trowbridge; "we shall have plenty to do today." Too soon day broke over nearby fields and woods where almost thirty-five thousand men had been killed or wounded or gone missing during the previous two days. Before midmorning the Fifth Michigan heard continuing sounds of war coming from Gettysburg, the barely muffled pounding of cannon accompanied by the clattering of small-arms fire.

These troopers didn't make it all the way to where most of the fighting would take place. Instead they took their posts with other regiments where two dusty roads crossed at the southern edge of a wheat field across from a ridge topped by woods. They were three miles east of Gettysburg, but the area they guarded was a route Confederates could use to attack the main Union Army's flank. Above them in the woods were cavalrymen commanded by J.E.B. Stuart, their adversary from four days before, in force with cannons that began firing down at them. Union rifled artillery shot back, eventually forcing the Confederate guns to withdraw.

The rebel First Virginia Cavalry did not pull back. Dismounted, they came forward to challenge the Union forces they outnumbered four to one. Because the Fifth Michigan had Spencer repeaters, its commanding officer, Col. Russell Alger, sent the men to meet the Virginians. After accepting the order, Major Ferry addressed the men under his command: "Now, boys, if any of you are unwilling to go *forward,* you may stay here." No one stayed. Their major could get them out of tight places, they felt. He knew when and where to take them. He'd pull them out again.

Leaving their horses behind, the Fifth Michigan advanced through the wheat with Ferry "all the while cheering, encouraging us on," a Michigan private remembered later, "and with our battery [artillery] in rear to overshell us, we pressed forward upon the enemy, forcing back their sharpshooters and battery."

With Ferry on the left and Trowbridge on the right, bullets flew around, one grazing Trowbridge's pants at the thigh, another striking an officer in front of him, another hitting a nearby rail, and another zipping past his head so close that he involuntarily dropped to the ground, leaving the man who fired at him believing he was dead.

A soldier felled by a Confederate bullet looked up at Ferry and said, "Major, I feel faint; I am going to die."

"Oh, I guess not," Ferry told him, "you are all right—only wounded in your arm."

Ferry picked up the man's Spencer and began firing at the Virginians before glancing along the line to shout, "Rally, boys! Rally for the fence!"

At that moment a bullet from a Virginian's rifle slammed through Ferry's skull, killing him instantly.

The Fifth Michigan, running low on Spencer cartridges, started to pull back while the Confederates resumed moving forward, this time supplemented by mounted cavalry. Seeing the outnumbered Fifth Michigan unable to return fire as before, the Seventh Michigan Cavalry charged onto the field with George Armstrong Custer in the lead, shouting, "Come on, you Wolverines!" heading straight to the mounted Virginians, driving them back up the hill. Stuart countered, sending the Michiganders into a disordered retreat, only to have Custer charge again, this time leading the First Michigan Cavalry "with his yellow locks flying and his long sabre brandishing through the air," remembered one cavalryman. "He looked like a fiend incarnate, the fire of battle burning in his eyes."

Fighting back and forth over the wheat field continued through a sweltering afternoon, the two Spencer-armed regiments—the Sixth and the resupplied Fifth Michigan—doing their parts. Noah Ferry's body remained where he had fallen.

Two hours before sundown, gentle southwesterly winds brought a summer storm to Gettysburg, its thunder meek compared to the sound of the day's artillery. By nightfall the three-day battle was over, its climax having been the slaughter of Confederate infantry in what would become known as the ill-fated Pickett's Charge. Where the Fifth Michigan fought and Noah Ferry died came to be called East Cavalry Field. Technically the fight there ended in a draw, which was really a win for the Union, because Stuart never got to the crossroads at the wheat field that could have brought him behind his enemy. For that he could place part of the blame on the Fifth Michigan Cavalry.

In his report eight weeks after the battle, Custer attributed the success of the Fifth Michigan "in a great measure to the fact that this regiment is armed with the Spencer repeating rifle, which in the hands of brave, determined men, like those composing the 5th Michigan Cavalry, is, in my estimation, the most effective fire-arm that our cavalry can adopt." Such a resounding endorsement, like the one Texas Ranger Jack Hays gave Colt's first revolvers after fighting Comanches in the summer of 1844, must have thrilled the Cheneys. New orders had already been arriving for Spencer rifles.

But not everyone on the Union side was a fan of Christopher Spencer's invention. While Custer was getting ready to write his report, Spencer was in Washington lobbying another prominent person, someone who didn't much care for his gun: Abraham Lincoln.

27

A LIVING SHEET OF FLAME

Lincoln's test of a Spencer wasn't going well. First the rifle's tubular magazine got stuck inside the gun's stock. Then, when it was finally liberated, it wouldn't work right. The president was no stranger to guns, and sometimes the high-up military brass was slow to modernize weaponry. If he liked a particular firearm, he would push the bureaucracy to get it. He hoped the repeater in his hands would be just such a weapon.

Lincoln picked up a second Spencer rifle, which worked well for the first two shots, but when he tried to load a third cartridge from the magazine, "the machinery somehow brought two cartridges forward in confusion, and so completely locked up the whole affair that we were unable to clear it out in less than a quarter of an hour," he wrote in a letter one month after Gettysburg. The substitute Spencer functioned as it was supposed to, once the tangled-up cartridges were cleared out, but the experience told the president that the gun was not combat-worthy.

In a letter to Maj. Gen. Stephen A. Hurlbut, who was eager to have repeaters issued to his men, Lincoln wrote:

> The result is that I have tried two of these guns; and each so got out of order as to have been entirely useless in a battle. This happened too, with specimens of the gun, which the Navy officers had inspected and bought for their own use, and, in fact, happened in the

> hands of their present Chief Ordnance Officer—The Secretary of War, for this, and for other general reasons, is opposed to furnishing this gun; and when to this is added the opposition of the Gen. in Chief, and my own discouragement, at the trials, I am sorry to disappoint you by saying I can not now order these guns for you.

It's unlikely that Lincoln was to blame for the mishaps with the two Spencers he had tested. His lifelong interest—passion, really—in how things worked helped him understand pretty much any machine and, if there was a way, to make it function better. "Lincoln had a quick comprehension of mechanical principles," observed his private secretary, John Hay, "and often detected a flaw in an invention which the contriver had overlooked."

Lincoln also liked new things. As a lawyer, he took on a number of patent cases, several involving agricultural machinery. These prompted courtroom battles over a cast-iron cemetery monument, mechanical reapers, improvements in plows, even a baby's cradle that rocked itself with a system of weights and pulleys. When the judge in the cradle case asked him how to stop it from rocking, Lincoln supposedly said, "It's like some of the glib talkers you and I know, Judge, it won't stop until it runs down." He also had his own ideas for inventions. One was for a ramming device run by steam for harbor defenses. Another was a steam-powered plow. These never went anywhere, but in 1849 Lincoln patented a system of inflatable bellows to lift boats higher in the water, so that they wouldn't run aground in shallows. Securing a patent was as far as Lincoln got. He never pushed to get his invention produced, though his hard work on the idea gave him the honor of being the only president ever to have patented something.

For Lincoln patents were not just a tinkerer's source of amusement or a way to make money as a lawyer. They were for people to reap rewards when they applied their brainpower to tasks in novel ways. Patents also enriched countries, especially the United States, a relatively new nation brimming with ideas for a more prosperous future. "In anciently inhabited countries," Lincoln said in lectures he gave at the end of the 1850s, "the dust of ages—a real downright old-fogyism—seems to settle upon, and smother the intellects and energies of man. It is in this view that I have mentioned the discovery of America as an event greatly favoring and facilitating useful discoveries and inventions." Before patent laws and their encouragement in the US Constitution, "any man might instantly use what another had invented; so that the inventor had no special advantage from his own invention. The patent system changed this; secured to the inventor, for a limited time, the exclusive use of his invention; and thereby added the fuel of *interest* to the *fire* of genius, in the discovery and production of new and useful things."

Chief of Ordnance General James Wolfe Ripley may have been hidebound about new weaponry, but his boss certainly wasn't. Lincoln nurtured a deep interest in guns, remembered one of his secretaries, "and had ideas of his own far in advance of some which were entertained by a few venerable gentlemen in the War Department." Inventors, both cranks and visionaries, usually "of some originality of character, not infrequently carried to eccentricity," found in the president an enthusiastic booster for new weapons. Many novel guns, like the Spencer, were brought to Lincoln personally. One oddity making its appearance at the White House was a rifle that could hit distant targets but had to be mounted on a spoked wheel as high as the shooter's shoulder. Lincoln didn't think much of it. Weird weapons kept piling up until his secretary's office "looked like a gunshop."

Lincoln would occasionally test guns on his visits with Dahlgren at the Washington Navy Yard; sometimes on a rubbish-strewn stretch of grass and weeds near the White House, where he used a pile of old building lumber as big as a small house for a backstop; and sometimes at the Washington Arsenal, where the Potomac and Anacostia Rivers met. One gun he found interesting was a rapid-shooting, multi-barreled weapon called a mitrailleuse, precursor to the Gatling gun, that was operated by a hand crank. The ammunition it used during the test lacked the power to penetrate a target set up for the demonstration at the Arsenal, so when Lincoln cranked the handle, the bullets bounced back around bystanders' shins. The president found this hilarious.

Lincoln seems not to have heard that Spencer rifles had given Colonel Wilder the edge at the Battle of Hoover's Gap, and Custer had yet to write his glowing report about how Spencers helped the Wolverines beat back J.E.B. Stuart in the Battle of East Cavalry Field at Gettysburg the month before. If he had been aware of its performance in battle, perhaps Lincoln wouldn't have dismissed the repeater so readily. There were other orders for Spencer rifles, so the Cheneys' company was keeping busy despite the president's unhappy experience with their gun. Having the commander in chief opposed to your product, however, was bad for business, especially if he was someone like Lincoln, who was intensely focused on weaponry for the Union. When word of the president's "discouragement" reached the Spencer company, treasurer Warren Fisher Jr., reached out to him. Perhaps Lincoln could be convinced that whatever went wrong that day could be explained or corrected. The inventor himself would be glad to come to Washington and "make a trial of the rifle in your presence," Fisher wrote. It would be good, he added, if the secretary of war; general in chief of all the Union armies, Henry W. Halleck; and anyone else who knew of the "mishaps of our gun at its former trials before you" could also be on hand for the shooting.

On August 17, four days after Fisher sent his letter to Lincoln, Christopher Spencer showed up at the White House carrying one of his repeating rifles in a cloth cover as a gift for the president. No guards greeted him before he was ushered into a reception room, where he found Lincoln alone. Spencer pulled off the cloth cover and handed the rifle to the president, who studied it carefully, as he did all mechanical things that intrigued him, especially guns that might be of use to the Union. The way he handled the repeater told Spencer that this was a man who knew firearms.

Please take it apart, asked Lincoln, so that he could "see the inwardness of the thing." After examining its workings, he told Spencer that the rifle's simple, substantial construction pleased him. Did the inventor have any plans for the next day? No, he was available if the president wanted him. "You come over tomorrow about two o'clock," Lincoln said, "and we will go down and see the thing shoot."

The following afternoon was cool and pleasant for mid-August in the nation's capital, certainly better than the sweltering days of the previous week when a number of people in Washington suffered sunstroke. Even the nights had been hot, but August 18 promised to be a comfortable day for shooting. Spencer was punctual and so was the presidential party, which included Lincoln's twenty-year-old son, Robert, and Charles Middleton, a Navy Department clerk whose company Lincoln enjoyed and who often visited the White House. Middleton took charge of the rifle and its ammunition, as well as a three-foot board with a black spot for a target at each end. On the way to one of Lincoln's favorite shooting locales near the White House, the group stopped in front of the War Department, where the president asked his son to see if Secretary of War Stanton wanted to join them. While they waited for Robert to return, Lincoln told stories and then noticed that a pocket of his black alpaca coat was torn. Taking a pin from his vest collar, he made a temporary repair, joking that it "seems to me that don't look quite right for the chief magistrate of this mighty republic." Stanton was too busy, Robert reported when he rejoined the shooting party. "Well," said Lincoln, "they do pretty much as they have a mind to over there."

After they set up the target, Spencer loaded seven cartridges into the rifle's magazine and handed the gun to the president. Lincoln's first shot was low and to the left, but the second scored a bull's eye with the other shots following close around it. "Now," he said, "we will let the inventor try it." The board was reversed, the bottom becoming the top, and Spencer took the gun. As familiar with his own rifle as anyone could be, he scored better than Lincoln. "You have beaten me a little," the president acknowledged when the shooting was over. "You are younger than I am and have a better eye and a steadier nerve."

Lincoln also shot the Spencer for an hour the next day. Whether those experiences improved the president's view of the rifle is not recorded, but it seems they did, as John Hay called Spencer's repeater a "wonderful gun, loading with absolutely contemptible simplicity and ease, with seven balls, and firing the whole, readily and deliberately, in less than half a minute." As for Spencer himself, he also got a nod from Lincoln's secretary, who referred to him as "a quiet little Yankee who sold himself in relentless slavery to his idea for six weary years before it was perfect. . . ."

Someone who didn't need to be sold on the Spencer rifle, of course, was Colonel Wilder. One month after Lincoln tried his hand at it in the company of the inventor near the White House, Wilder and his Lightning Brigade summoned its firepower again, this time near a creek called Chickamauga that meandered through forests and farm fields in the mountains of northwestern Georgia.

By mid-September, General Rosecrans's Army of the Cumberland had taken control of Chattanooga and north Tennessee. The Confederates under General Braxton Bragg wanted to take them back, especially Chattanooga, with its crucial railroad hub, so they decided to lure Rosecrans into thinking he could defeat a smaller, retreating Southern force to the south. Only the force wasn't small—sixty-five thousand men, compared to Rosecrans's sixty thousand—thanks to the arrival of large reinforcements from Virginia and Mississippi, and it was ready for him. If Bragg could cut the Army of the Cumberland's supply lines from the north, he would then try to destroy it while reclaiming precious Chattanooga. If the Union line was spread thin enough—and it would be very thin over difficult terrain—it appeared the Confederates could do just that.

The Lightning Brigade's five regiments were only a small part of the Union forces when the battle began, but their Spencer rifles made them seem like an entire army. Horses gave the Union soldiers mobility to dash where they were needed, which sometimes seemed everywhere, as Confederate attacks on Union lines were often piecemeal and uncoordinated. The men were well supplied with ammunition; each had sixty cartridges in his knapsack and a hundred more in a bag attached to his saddle. Early in the two-day main battle—they actually started skirmishing the day before—a mass of Confederates left cover to cross a small field, while Wilder's men waited in woods on the other side. As soon as the rebels were fully exposed, the Spencers unleashed a continuous blast of fire, with Capt. Eli Lilly also shooting at them with cannons loaded with canister, turning them into big shotguns. "Our entire line from right to left opened out upon them with Artillery and small arms," remembered a corporal in the Seventeenth Indiana Mounted Infantry. "The rapid firing of our seven shooters gave our line the appearance of a living sheet of flame, while we knelt down in two ranks behind our works."

A private in the Seventeenth Indiana wrote, in his account of the battle:

> Again and again, all day each side charged back and forth across the field. Often we could not see a dozen yards in front of us for the smoke, but never did the enemy once reach us. We held our line intact. Many times during the day we had to renew our supplies of ammunition. We shot so much that before night each man had a little pile of empty shells, and I remember one time that several of the soldiers near me in a lull of the fighting compared piles to see which one had done the most shooting.

"The effect was awful," reported *The Indianapolis Daily Journal.* "Every shot seemed to tell. The head of the column, as it was pushed on by those behind, appeared to melt away or sink into the earth, for though continually moving it got no nearer. It broke at last and fell back in great disorder."

But the Confederates rallied, pushing through the gunfire until they dove into a ditch near the Union line for shelter. It didn't save them. Lilly whirled two cannons around and blasted canister into the ditch, shredding the soldiers bunched there. "At this point," Wilder said, "it actually seemed a pity to kill men so. They fell in heaps, and I had it in my heart to order the firing to cease to end the awful sight."

On the battle's second day, Wilder's troopers let loose on General James Longstreet's advancing soldiers with such force that for a moment Longstreet thought he was facing an entire corps.

A soldier in the Thirty-ninth Indiana Mounted Infantry recalled:

> At a distance of less than fifty yards six solid lines of gray were coming with their hats down, their bayonets at a charge, and the old familiar rebel yell. Our first volley did not check their ad-

> vance, but as volley after volley from our Spencer rifles followed, with scarce a second's intermission, and regiment after regiment came on left into line on our right, and poured the same steady, deadly fire into their fast-thinning ranks, they broke and fled.

A sergeant in the Seventy-second Indiana Volunteer Infantry believed "such slaughter and carnage as our Spencers worked would surely delight the worst demons in Hades."

In the end the Battle of Chickamauga was a Confederate victory, though the Union rout would have been worse had Wilder's Lightning Brigade not poured Spencer fire on the soldiers of the South, slowing them down, breaking their lines, and allowing the Army of the Cumberland to make its way back to Chattanooga. "How I wish our infantry could have all been armed with Spencer rifles today," lamented one Union private, and "all the force of Rebels in Dixie could not have forced their way through our lines."

"We got through the three days of fighting without losing many men," Maj. James A. Connolly of the 123rd Illinois Volunteers told his wife in a letter a month after the battle. The Lightning Brigade was lucky in that. More than twenty-eight thousand other soldiers, Union and Confederate, were not so fortunate, having been killed or wounded in the bloodiest battle of the war outside of Gettysburg. "We think our Spencers saved us," Connolly wrote, "and our men adore them as the heathen do their idols."

28

NO BETTER FRIEND, NO WORSE ENEMY

As the gunmakers' markets matured through the Civil War era, some began mastering the art of product promotion, following the lead set by Samuel Colt. For Oliver Winchester that promotion came from his own fan base on the front lines. Among his enthusiasts was Capt. James M. Wilson of the Twelfth Kentucky Cavalry, who said he owed his life to a Henry rifle.

Living in a Confederate-leaning part of Kentucky in 1862 was risky for a solid Union man like Wilson, so when neighbors threatened to kill him, he stashed a Henry rifle and a Colt revolver in a log crib not far from his front door, just in case they tried. One day they did. Seven armed guerrillas burst into the Wilson home while his family sat at the dinner table, shooting on their way in, shattering a glass of water in his wife's hand, but hitting no one. "For God's sake," Wilson shouted, "if you wish to murder me, do not do it at my own table in presence of my family." All right, the intruders said, step outside.

The moment he reached the front door, Wilson leapt for his hidden guns, while bullets tore through his clothing without touching flesh. He grabbed his Henry rifle and fired five shots, each one killing a man who was shooting at him. The other two guerrillas then sprang for their horses. As one grabbed the pommel of his saddle, a bullet from Wilson's repeater blew off four of his fingers, but the man was able to mount his horse anyway and was trying to flee when another bullet toppled him to the ground, dead. Wilson's eighth shot killed the seventh man.

Word of Wilson's shoot-out reached a delighted Oliver Winchester, who wrote him on New Year's Eve 1862:

> A few days since I heard a detailed and thrilling account of your adventures with Guerillas, in which your coolness and courage were conspicuous, aided by a skilful use of Henry's Rifle. I do not know how true the account is, but feel a great interest in obtaining a reliable statement of any facts showing the efficiency of our arm in competent hands. . . . I trust I am not asking too much of you, and will esteem your compliance a great favor, which I shall be happy to reciprocate in any way in my power.

Seven weeks after it was sent, Wilson finally received Winchester's letter. Yes, he told the gunmaker in a letter from Kentucky, he did use a Henry against seven guerrillas, finding it "particularly useful" in the number of shots he could fire and "its fatal precision." As far as he was concerned, Wilson wrote, "the Henry Rifle is decidedly the *best* gun in the service of the United States." Give him "sixty men armed with the Henry Repeating Rifle, with a sufficient quantity of cartridges," he claimed, "and it is not an overestimate to say that we are equal to a full regiment of men armed with muskets."

Winchester harvested more testimonials, including one from Maj. Joel Cloudman of the First District of Columbia Cavalry. Cloudman and several fellow troopers were captured, along with their Henry rifles, and spent some time in the Confederacy's infamous Libby Prison. There he heard a rebel say, "Give us anything but your d__d Yankee Rifle that can be loaded Sunday and fired all the week." And the president of the strategically located Louisville and Nashville Railroad asked Secretary of War Stanton to "supply him with 300 repeating rifles, those of Henry preferred," to protect trains from marauding Confederates. "The increase of guerrilla bands has been such that unless those engaged in running the trains are armed it will not be possible much longer to retain them in the service."

Other accolades came in letters to newspapers from a sender identified only as "L.W.W.," who warned that Henry rifles in Confederate hands, which happened on occasion, "could cause terrific losses among gallant fighting men, causing this conflict to drag on eternally, and cause loss of spirit." Already there was discontent in the ranks, he said, "and if these terrible weapons fall into the wrong hands, the hurt will be too great to bear. Our boys, who are willing to sacrifice all, must be provided with the latest and most modern arms. Nothing else will satisfy the Union and all loyal citizens." L.W.W.'s enthusiasm for the Henry rifle is easily explained. He was Oliver Winchester hiding his identity.

The prospect of being armed with a repeating rifle was so appealing it was used to lure men into enlisting. In March 1863, three weeks after Congress passed the first wartime draft in the nation's history, Capt. C. A. Barton placed an advertisement in an Ohio newspaper with the heading "An Easy Way to Avoid the Conscription." Join his company of sharpshooters, Barton promised, and you would receive a bounty. What's more, "We also can give you Spencer's Repeating Rifle, the most effective weapon now in use."

The Spencer also had a champion in Union Brig. Gen. Edward W. Hinks. At the end of April 1864, Hinks had a specific reason for wanting his troops armed with them. On April 12—the third anniversary of the first shots fired on Fort Sumter—Confederate cavalry under Maj. Gen. Nathan Bedford Forrest had overwhelmed a vastly outnumbered Union garrison at Fort Pillow, Tennessee, easily forcing the Federal troops there into submission. The battle itself wouldn't be remembered as much as its aftermath would.

What followed the fighting was called a massacre, with Confederate soldiers murdering captured Union men, many of them wounded. African American prisoners fared worse than their white comrades. Many black soldiers, seeing harsh treatment being meted out, jumped into the river to escape, only to be shot at by Confederate soldiers on the bluff above. "From where I fell wounded," remembered Union Lt. Mack Leaming, "I could plainly see this firing and note the bullets striking the water around the black heads of the soldiers, until suddenly the muddy current became red and I saw another life sacrificed in the cause of the Union." Leaming then saw a pair of Confederates pull a black Union soldier from the river. "He seemed to be wounded and crawled on his hands and knees. Finely [*sic*] one of the confederate soldiers placed his revolver to the head of the colored soldier and killed him." A mate on a Navy steamer who came to Fort Pillow the day after the battle saw the bodies of several African American soldiers "with their eyes punched out with bayonets; many of them were shot twice and bayonetted also."

Seventy percent of the white soldiers survived the attack on Fort Pillow and its aftermath, while only thirty-five percent of the black soldiers did. The brutality of Southern soldiers toward armed men they thought should remain slaves did not go unnoticed beyond Fort Pillow. One who noticed was General Hinks, who commanded a Union division of African American troops in Virginia. Two weeks after the Fort Pillow massacre, he wrote to Maj. Gen. Benjamin F. Butler, the commanding general of the military department covering Virginia and North Carolina, and asked that his soldiers "be more efficiently armed, to enable them to defend themselves and lessen their liability to capture" in light of "the recent inhumanities of the enemy perpetrated upon troops of like character to those of my command."

It was often the lot of African American soldiers to be given inferior weapons, and the men under Hinks were no exception. Unreliable old guns might be fine "for troops who will be well cared for if they fall into [enemy] hands," he told Butler, "but to troops who cannot afford to be beaten, and will not be taken, the best arm should be given that the country can afford." The right thing to do, Hinks insisted, "is to arm our colored troops with Spencer repeating rifles, and I request that my division, or a part of them, may be armed with a repeating or breech-loading fire-arm."

Naturally business was good for many gun companies during the war. The Springfield Armory could not meet all the demand, and a repeating cartridge rifle was not something the armory undertook to produce. The military ordered so many Spencer rifles in 1864 that the Cheneys had to contract with the Burnside Rifle Company in Providence, Rhode Island, to meet Ordnance Department orders. That firm's product was a single-shot, breech-loading carbine invented and patented by Ambrose E. Burnside, who became a not terribly successful major general in the Union Army. His invention was a hit with the military, though it was on its way to obsolescence as the Civil War approached its end. When Oliver Winchester learned that the Burnside firm was a defendant in a patent suit brought by an India rubber magnate, he decided to fight his competition by secretly buying the patent that was the subject of the litigation. When the patent was eventually declared valid, Winchester demanded that damages be paid in the form of machinery, a move that forced the Burnside company to stop making guns.

Henry rifles were still coming from the New Haven Arms Company in relatively small numbers (between 250 and 350 a month in 1864). Part of the blame for slow production belonged to inventor and now shop superintendent B. Tyler Henry, who never increased the workforce, even when orders for rifles rose and the company bought extra machinery to handle the demand. Although Henry kept things going, increasing market share seems not to have been among his priorities. This hurt business and angered Oliver Winchester, who leased an armory in Bridgeport on his own to expand Henry rifle production. Digging into his personal finances, Winchester bought enough equipment to double the number of rifles the company could produce, but he had one problem: Henry had an exclusive contract to make guns for the New Haven Arms Company into June 1864. Until that moment arrived, Winchester's leased Bridgeport factory had to remain idle, despite the country's war needs.

A fire in February 1864 destroyed much of the Colt factory—arson by Confederate saboteurs was considered a likely culprit, though the fire was probably an accident—but widow Elizabeth Colt ordered it quickly rebuilt to continue making arms for the Union. Remington and Colt had been making tens of thousands of both long guns and revolvers for the war.

Smith & Wesson was never a contender for government sales during the war, so the partners did not have to depend on government contracts. They did well anyway with their low-powered revolvers. So many Union officers, not to mention civilians, liked them enough to put their own money into buying them as extra protection on the battlefield. Smith & Wesson also had a lock on the bored-through cylinder for at least a few more years and let other makers use it—for a fee, of course. And, also of course, some gunmakers decided to get in on the act without paying Smith & Wesson anything. One who took that route was Horace Smith's old boss from his pepperbox-pistol days, who had begun making a revolver of his own. Ethan Allen and his brother-in-law and partner Thomas Wheelock (Allen's first brother-in-law-partner Charles Thurber had retired by then) staked out a position very much like the one the Massachusetts Arms Company took when it fought Samuel Colt over the Edwin Wesson patent in 1851. They said that Rollin White's bored-through cylinder wasn't original after all, and therefore they couldn't be barred from making their own version. In 1859 White, along with Smith & Wesson, filed a federal lawsuit in Massachusetts against Allen, though the case didn't come to trial until 1863.

Thus, in the middle of the Civil War, with the federal government desperately in need of modern weaponry, two gunmakers fought a court battle over who had the right to manufacture the most advanced revolver in the country, if not the world. As in the Colt suit against the Massachusetts Arms Company, the opposing sides hired high-level lawyers, including Allen attorney Benjamin Curtis, who had been a Supreme Court justice until he resigned after filing a fiery dissent in the infamous *Dred Scott* case. Also as in the earlier patent fight other firearms companies were watching. Some wanted to produce big-bore revolvers like Colt's and Remington's but using metallic cartridges, rather than the slow-to-load, ball-powder-and-cap combinations, but they couldn't do so without infringing on the White patent. Smith & Wesson considered introducing a .44-caliber cartridge revolver under contract with the Whitney Armory, but they backed off after Rollin White angrily insisted that his agreement with them forbade the partners from making a deal with Whitney.

Smith & Wesson accused Allen & Wheelock of selling twenty-five thousand pistols in violation of White's patent and depriving them of about $50,000 in profits. That money should be paid to Smith & Wesson, its lawyers argued, and any infringing pistols must be either turned over to the firm or destroyed. Allen & Wheelock's lawyers countered as the Massachusetts Arms Company had in its fight against Colt: There was nothing new in White's "pretended invention" of a bored-through revolving cylinder. Examples of previous patents could be found in Europe and the United States, a few of which the lawyers brought to court. White's creations, according to Allen's lawyers, "were experimental, unsuccessful and fruitless, and were abandoned," while the other inventors diligently worked on making something worthy and new.

In November 1863 Smith & Wesson—and Rollin White, of course—triumphed. Deciding the case was Supreme Court Justice Nathan Clifford, doing duty as a trial judge. The man Clifford replaced on the high court was Benjamin Curtis, who was now Allen's attorney. That odd situation meant that Clifford was presiding over a case in which the lawyer for one side had been his predecessor on the bench. In a long and windy opinion, Clifford accepted as true everything White had said about how he experimented and why he had delayed seeking patent protection. And the other patents, Clifford decided, did not invalidate White's. Allen appealed, but that didn't stop the message from going out to other gunmakers, as it did when Colt won his 1851 patent case: Dare make a gun that infringes on our patent, and we will crush you.

Then on April 9, 1865, something happened that changed everything in America's gunmaking world. Robert E. Lee surrendered his Army of Northern Virginia to Ulysses S. Grant. The Civil War was clearly ending. The federal government, once starved for guns, now had too many. Less than three weeks after Lee's surrender, the War Department ordered the chief of ordnance to "stop all purchases of arms, ammunition, and materials therefor, and reduce the manufacturing of arms and ordnance stores in the Government arsenals as rapidly as can be done without injury to the service."

With that the gunmakers' gravy train came to a halt—except for Smith

& Wesson. Demand was so heavy, even without government contracts, that Smith & Wesson had two years' worth of orders waiting to be filled. Edwin Wesson's rifle business might have been saved by a big government contract back in the late 1840s, but brother Daniel and senior partner Horace Smith were finally on solid financial ground without one. They were now the two wealthiest men in Springfield. Their small revolvers, able to load state-of-the-art metal cartridges, continued to sell briskly. In 1865 alone the firm sold 55,543 of them. Other gun companies sold thousands more with the White patent, paying a fee to Smith & Wesson for the privilege. Rollin White launched his own revolver-making firm under his name. He, too, had to pay Smith & Wesson, since the partners had the license for the cylinder White had patented. The arrangement Smith & Wesson had made with Rollin White was a gold mine for the firm, one that promised to keep producing wealth as long as all remained happy with each other. But the bonds between Smith & Wesson and White were fraying. Almost from the beginning the inventor felt the financial pinch of pursuing infringers at his own expense, while Smith and Wesson were enriching themselves. White nursed his grudge, looking for a way to even the score.

Daniel Baird Wesson held on to his brother's order book, the one Edwin had opened with hope and a touch of humor and Daniel had closed with a plaintive "Thus ended the manufacture of Rifles by the far famed E Wesson." Now, with the Civil War over and Smith & Wesson a success, Daniel pulled out the old book. On October 16, 1865, beneath his long-ago lament, he wrote a second message, also to no one in particular: "Very great are the changes which have taken place since the day on which the manufacture of E Wesson's Rifles seased [*sic*] now nearly twenty years."

29

THE WAGES OF PEACE

As it mourned the loss of 620,000 men under arms, the American nation set about renegotiating its relationship with its own advancing weaponry. The gun barons of the day had to adapt to a sudden slackening in the demand for their wares.

When the shooting was over, "the most fantastic conglomeration of assorted small arms ever to be collected together on the face of the earth" lay "stacked in piles like cordwood, or in armory chests of twenty muskets, twenty bayonets, with appendages," one firearms historian wrote. There was no work for them to do, the Confederacy having surrendered, so they rested quietly in the war's wake.

"[T]he armories are disbanding their forces," *Scientific American* observed in late June 1865, "the makers of ordnance are unemployed, and the whole tenor and tone of our daily lives is as suddenly transformed from one of eager and vigilant activity for our national existence as if we had dropped from one sphere to another."

Private arms makers produced far more guns for use in the Civil War than did the Springfield Armory. Without a network of private companies with interchangeable-parts experience, the government would never have met its wartime needs. And without those private companies, repeating firearms would not have become part of the nation's arsenal as soon as they did. But to make small arms in abundance, gunmakers had to invest heavily in new equipment, which was fine when the government wanted as much up-to-date weaponry as it could buy. Now the future for many gun companies was uncertain.

Machinists who helped the gun companies succeed took their talent to other enterprises. A former supervisor at Robbins & Lawrence who helped the Providence Tool Company gear up to make rifled muskets, left weaponry to work on sewing machines. A Colt machinery contractor also went into sewing machines, eventually becoming president of the Weed Sewing Machine Company. Another key worker for Colt and Remington during the war started a firm with Christopher Spencer that would pioneer drop forging. Men like these, and the small-arms firms they had been a part of, were critical to the development of machine tools, interchangeable parts, and precision measurement throughout American industry.

In the five years following the Civil War, twenty-seven New England gunmaking firms closed their doors, while only eight launched new businesses. For the companies that survived, the postwar years were a period of retrenchment. Salvation for some came from expanding into overseas markets, where governments were eager to take advantage of American firearm innovation. In the United States the cross-continent movement picked up speed, with emigrants arming themselves and the Army turning its focus westward. The civilian market began taking on more prominence.

For a couple of major gun companies, peace wasn't the only problem. For them tension between inventors and employers threatened to undermine business even further.

Before the Civil War's end, Oliver Winchester was already looking abroad for Henry rifle customers. If large orders weren't coming from the US military, as he saw it, perhaps other countries preparing for conflict would want to buy the latest in modern weaponry. Winchester was right. Just such a place was the Kingdom of Bavaria, which made a deal with Winchester for five hundred Henrys fitted with improved loading mechanisms. Since inventor B. Tyler Henry's exclusive contract had ended, Winchester opted to use his rented Bridgeport factory to make the kingdom's guns starting in late 1864. By then the clothing operation of Winchester & Davies was doing so well that Oliver Winchester and his partner tapped their sons to run it. This freed Winchester to focus on guns, which he had finally learned a lot about, at least enough to come up with his own ideas to improve the Henry rifle. Sensing European markets beyond Bavaria, Winchester boarded a ship bound for Naples, Italy, arriving there in mid-January 1865.

Though he was no longer a barrier to large-scale production, Henry—the golden inventor whose name identified the rifle Oliver Winchester made—was still an irritant. There was no doubt he was a hard worker and a near-genius when it came to mechanics. Before Henry came along and added his improvements, the Smith-Jennings rifle and the Volcanic did nothing more than spit underpowered bullets and gobble up investors' money. Because the Henry rifle was a serious weapon, some Union soldiers deployed it in battle, allowing Oliver Winchester to feed his personal fortune. But Henry was no businessman. His sluggishness in making as many rifles as Winchester insisted on had never improved, causing ongoing tension between inventor and owner.

Before departing for Europe, Winchester gave company secretary Charles W. Nott a power of attorney to conduct routine business matters while he was away. This detail did not escape the attention of B. Tyler Henry, who saw an opportunity to advance his own frustrated aspirations. Not only was he the mechanical wizard whose mind gave the Henry rifle to the world, he owned company stock. With Winchester an ocean away, Henry persuaded Nott and some key New Haven Arms Company stockholders that it was time to recharter the firm, change its name to the Henry Repeating Rifle Company, and put him in charge. And why shouldn't it be called the *Henry* company? After all, everyone, including the New Haven Arms Company, called the guns *Henry* rifles. *His* name was on the lips of admiring soldiers, and *his* initial "H" was on every cartridge loaded into the only rifle the New Haven Arms Company made. Henry also bore a grudge—his arrangement with Winchester didn't allow him to reap royalties from his invention. Who was Winchester to get richer while Henry couldn't profit from his own genius? The disgruntled inventor and those he had enlisted in his cause then petitioned the Connecticut State Legislature to do what they wanted.

If Henry and others thought Oliver Winchester would quietly accept what they had done, they didn't know him well enough. Word reached Winchester in May 1865, while he and Jane were in Switzerland preparing for a round of visits, which included enjoying the comforts of a few European spas as well as pursuing foreign contracts for Henry rifles. An enraged Winchester immediately canceled his plans and booked a return to New Haven. At the same time, he cabled his London bankers to instruct Davies back in New Haven to "present all those mortgages and liens now in my possession or held by my son, William Winchester, against the company formerly called the New Haven Arms Company, to its successor in business, the Henry Rifle Company [*sic*] for immediate collection."

The company's debts to Winchester personally were big enough to force the renamed, rechartered firm into bankruptcy if he applied full pressure for repayment. By the time he arrived back in the United States, Winchester had chosen a better strategy. Rather than squeeze the company to death or challenge the firm's rechartering, he would set up his own company to make improved Henrys. The Bridgeport factory was Winchester's to command, which meant not only that it could make guns for him but that he could refuse to let it make guns for the Henry Repeating Rifle Company. In itself this move would cut the rival firm's production capacity in half. Using leverage from his own stock holdings as well as his son's and that of the Davies father and son, Winchester arranged to have himself elected president of the new Henry firm, booting Nott and replacing him as secretary with a Winchester loyalist. Soon he would control two gunmaking enterprises, one of them the new Winchester Repeating Arms Company based in Bridgeport. He could then orchestrate folding what had been the New Haven Arms Company into the gun business bearing his name.

The year 1866 brought both triumph and tragedy for Oliver Winchester. That spring the first rifle to be called a Winchester was born. The Winchester '66, a much improved Henry rifle, suited travelers going west. It was more reliable than the Henry, with a magazine less susceptible to malfunction when exposed to mud and grime. It was also significantly easier to fill with cartridges, because it had a loading port mounted on the side of its frame. It also appealed to the civilian market, which the company needed if it was to profit from peace. A brass receiver holding the mechanism earned the Winchester '66 the colorful nickname "Yellow Boy."

It was also the year that Oliver Winchester acquired a public title. Ever since handing out William Henry Harrison badges at his clothing shop during his Whig days in Baltimore, Winchester had dabbled in politics. He had been a New Haven city councilman and a member of the city water committee in 1853, and when one of the six Connecticut electors for Abraham Lincoln in 1864 died of typhoid fever, Winchester was quickly chosen to replace him. Now, two years later, Civil War hero Gen. Joseph R. "Fighting Joe" Hawley wanted Winchester to join his ticket when he ran for governor of Connecticut, a state whose Republican party was split between Radicals—who urged harsh reconstruction of the conquered South with suffrage for blacks—and conservatives, who wanted leniency toward the defeated Confederacy along the lines Lincoln had advocated. Also in play was a rivalry between more conservative, industrial New Haven and the more radical financial and insurance factions in Harford.

The Democratic opposition was unified behind James E. English, a wealthy self-made New Haven clockmaker, and if Republicans wanted to win the 1866 gubernatorial election, they needed to build bridges between their party's factions. Hawley, a Hartford newspaperman long aligned with abolitionists, could use a prominent New Haven industrialist in his camp to achieve that, someone who was tough and bold and not a confirmed Radical. Oliver Winchester was such a candidate. Winchester's wealth and his readiness "to respond liberally to calls made for party purposes" sealed his nomination for lieutenant governor, according to a newspaper account at the time. "The manner in which Mr. Winchester received the nomination," *The Boston Post* reported, "was a striking example of the old adage 'money makes the mare go.'" Preference for other candidates at the convention faded as attendees picked "the man who had the greatest abundance of lucre. And from all accounts, Mr. Winchester has not abused the confidence placed in him, but has bled freely and submits to it with the best possible grace."

Hawley went on to win the election for governor by a narrow margin of 541 of the 87,417 votes cast. Winchester fared a little better, with an 872-vote victory over his opponent. Connecticut governors at the time served one-year terms, and very soon after he was elected, Winchester decided he would not run again. It appears he did little officially during his year as lieutenant governor, spending some of his time in office visiting Europe to sell his rifles and performing duties as one of the thirty members of the American commission to an international exposition in Paris.

Winchester didn't hide his political leanings while he was lieutenant governor. Shortly before he took office, he and the just-elected Connecticut secretary of state received a scathing rebuke from The *Hartford Daily Post* for joining conservatives whose Southern "sympathies and affinities may be very effectually disguised," an action that "cannot but be heartily regretted by every lover and advocate of true Union principles." Those two men, the *Post* continued, "have taken a step which will ultimately cover them with merited disgrace."

As summer began, 1866 promised to be Oliver Winchester's most glorious year so far. The fatherless boy from a cash-strapped childhood was now one of Connecticut's richest citizens, a man who had just been placed one step away from the governorship. His name was known outside the state and well beyond the country's borders. He had been the conqueror in a corporate war that swelled his already impressive wealth, while the prospect of real war elsewhere gave him hope for more. Perhaps best of all, the wife of his only son and heir to his businesses was about to give birth to the next member of the new Winchester industrial dynasty.

The hoped-for heir apparent arrived on June 15, although the baby turned out to be a girl, not a boy. It was soon clear that eating was difficult for tiny Annie Pardee Winchester. Something was amiss in her digestive system. Nothing the Winchesters and their doctor tried could slow the baby's steady weight loss. The cause, the doctor concluded, was marasmus, a form of malnutrition usually found in children consigned to dire poverty. The Winchester family lacked for nothing, so failure to provide proper sustenance or care could not be blamed. They brought in a wet nurse, but Annie continued to waste away, starving slowly and painfully until July 24, when she died. Two days later the Winchesters buried her in New Haven's Evergreen Cemetery, near her aunt Ann Rebecca, who had married a Presbyterian minister, and Ann's infant son, both of whom had died two and a half years earlier.

In early 1867 three prominent pro-Johnson Republicans urged Winchester to change his mind and run for reelection despite opposition from Radicals, who accused him of being a traitor to the cause. Winchester wouldn't budge. Besides, he told them, "my business relations have assumed such a form as to require my undivided attention, and may, at any moment, call me out of the States, and keep me away most of the ensuing year." Then he unloaded on the Radicals opposed to President Andrew Johnson's moderate treatment of the conquered South. In a quickly published letter he wrote:

> The audacity with which they press upon Congress and the country the most revolutionary doctrines and disorganizing measures, is truly alarming. . . . The only excuse offered for this abandonment of principle is *progress*. It is a fearfully declining progress, which, if not soon arrested, threatens us with political anarchy, and financial and social ruin. Believing this, I cannot but feel that there is no evil or danger so great as this downward progress, and that it is my duty to join with all good and true men in arresting it, and aid in restoring the government to men who will legislate wisely for the best interests of the country. . . .

Winchester then threw his support to Democratic candidates, including James English, who was again running against Hawley.

As in the previous year's contest, voters who turned out for the 1867 election were closely divided, though this time they gave English the victory. A few months after leaving the governor's office, Hawley bought *The Hartford Daily Courant* and started taking on new public duties, eventually serving as a United States senator for four terms. For Oliver Winchester 1867 marked the end of his pursuit of elective office. Nevertheless, if Samuel Colt could insist on being called "Colonel," Winchester had the right to be called "Governor," and he was for the rest of his life.

Winchester's main competitor in the Civil War, the Spencer Repeating Rifle Company, was still operating in 1866, but its future was bleak. The military was awash in Spencers, and though the cavalry could use them against Indians on the western frontier, there was no need for more. If the postwar squeeze on firearms manufacturers drove the Spencer firm into extinction, other companies might benefit, provided the need for a new wave of repeaters like the Spencer arrived. And if so, Oliver Winchester would be waiting.

The United States government was on the hook for the Spencer carbines it had ordered in late 1864, and it paid for them, even though the first batch arrived as the war was beginning to wind down and six days after Robert E. Lee surrendered his Army of Northern Virginia to Ulysses S. Grant. Spencer rifles and carbines were everywhere then. Even Abraham Lincoln's assassin, John Wilkes Booth, had one with him when he was hunted down and shot in a Virginia tobacco barn.

That oversupply was one of the Cheneys' problems. The Spencer company had plenty of them on hand too. In its 1866 catalogue the firm boasted, "In *range* and *force* the Spencer Rifle is second to no other arm. It will throw a ball *two thousand yards*, and may be relied upon for general accuracy at a greater distance than any other arm yet invented." In addition to that dubious puffery, the catalogue included accolades from several military luminaries, including Colonel Wilder, Governor Hawley while he was in the Army, Gideon Welles, William Tecumseh Sherman, and George Armstrong Custer. Company treasurer Warren Fisher Jr., had asked Ulysses S. Grant what he thought of the Spencer. Grant's response was too good to be excluded from promotional material: "The war has nearly settled the question that the Repeating Arms are ultimately to take the place of all others, and, so far, I believe none stand higher than the Spencer."

The inventor himself had nothing to do with running the business other than a brief stint as plant supervisor. He had collected his dollar-a-gun royalty until 1863, when he dropped it to fifty cents. Machines were his specialty, including the machines he designed to make the company's rifles. When the firm no longer needed him to be its lead salesman, Christopher Spencer was free to embark on other ventures, which he quickly did. In early 1866 he left Boston for Roxbury, where he joined Sylvester Roper—another inventor of many things, including firearms—to make Roper's patented repeating guns. It was up to the Cheneys and others whose money was tied up in the Spencer company to salvage what they could of their business. They continued to tout the Spencer's fast-firing capabilities, of course. On the first page of their 1866 catalogue, they boasted that a rifleman could get off seven shots in less than half the time it took to load and fire a single-shot musket. "More rapid firing than this," the catalogue cautioned, "even if attainable, would be wholly undesirable."

For the civilian shooter who wanted to take careful aim and still have cartridges in reserve, the company made sporting rifles with fine American walnut burl stocks, hoping they would lift its fortunes. Unfortunately for the Cheneys, they did not. There wasn't enough demand, though the company made a custom sporting rifle for celebrated cavalry hero Custer after the war. When the Winchester Repeating Arms Company heard that the Spencer firm was sliding toward bankruptcy in late 1868, its board of directors authorized Oliver Winchester to explore buying its Civil War rival. A look at the Spencer company's books told Winchester the business wasn't worth bidding on. About a month later a fledgling firm named Fogarty Repeating Rifle Company took a gamble and bought all the Spencer assets, soon forming the American Repeating Rifle Company. At that moment the Spencer company as a separate entity ceased to exist, having produced its signature repeaters for less than seven years.

But it turned out that Oliver Winchester wasn't done with the Spencer firm. Within a year after buying Spencer, the American Repeating

Rifle Company went bankrupt. The Winchester Repeating Arms Company quickly circled back and bought all the assets—guns, machinery, and patents—for $200,000. Winchester held on to the Spencer rifles and patents, but on September 28, 1869, he auctioned off all the equipment, including lathes, drills, more than 150 milling machines, and a sixty-horsepower steam engine, for a total of $138,000. With this move, Winchester acquired patents that could have been used to compete with his guns in the future, plus many thousands of Spencer repeaters ready for sale to customers interested in fighting wars.

By this time Christopher Spencer was back in Hartford with a new business partner making an assortment of rifles and shotguns. The Hartford area was his true home, where his family had long lived. Two months after war broke out, he had wed a local girl there, twenty-four-year-old Frances Theodora Peck, called Dora, daughter of farmer and laborer Sardis Peck. So far he and Dora had no children, which was probably a good thing, since she needed to help care for her father, whose mental state had been deteriorating for years. Since 1865 he had suffered from "mania," a term usually applied to someone who was wildly overactive. After Spencer returned from Amherst, Dora and her divorced older sister were living with their parents in Hartford. Sardis Peck's eroding sanity soon led to his confinement in an asylum, where he would sometimes be held down with straps. He would never recover.

Affable salesman and brother Samuel Remington was in the right place for the lavish living he enjoyed more than his strict, business-minded brothers did. The late 1860s was also the right time to be in Paris, swirling about in the company of moneyed people and other higher-ups, his radiantly beautiful wife, Flora, on his arm at elegant balls and soirées attended by all the right sort. Streets in nighttime Paris glowed, illuminated by tens of thousands of gas lamps that justified the French capital's nickname, City of Light. Of course Samuel's presence there was not just to have a good time. He had a gun to sell, and he didn't have to overstate its virtues to convince buyers.

The Remington firm had done well in the Civil War, selling enough revolvers to rank it second in military customers behind Colt. For gun companies in general, war's end meant loss of revenue, idle equipment, and sometimes debt. Remington was no exception. Guns were the brothers' main business, so they felt the postwar squeeze even though the Remingtons' product line had included agricultural implements well before the war began. They also had an employee who came up with a weapon that would improve the company's fortunes. This firearm gave Remington a needed lift. It was a single-shot rifle, not a fast-shooting repeater, but its design was strong, simple, easy to use, and less susceptible to harsh treatment than many other weapons. Later dubbed the "rolling block" for its thick disk that rolled down behind the chamber to admit a cartridge, its mechanism had already been reviewed favorably by US Navy Ordnance in a pistol. Samuel Remington took several of the rifles with him to Paris.

Before Samuel and Flora left for Europe in the summer of 1866, he and Philo switched roles at the company; Philo turned the presidency over to vice president Samuel, who then became head of the firm. They did this to give Samuel an extra measure of prestige when negotiating deals abroad. And he was unquestionably the right man for the job. If he wasn't as brazenly theatrical as Samuel Colt, Samuel Remington, with his dash and welcoming green-blue eyes, was every bit as charming. His captivating personality and the rolling block made an unbeatable pair. Remington toured Europe with his rifles, pushing them as the new marvel in reliable power on the battlefield. Sweden liked them enough to place an order. So did Denmark. The real boost came at the International Exposition of 1867 in Paris, when the rolling block won a silver medal, the highest honor for a rifle. (Another medal recipient was the Spencer repeater, though it did little good for the Cheneys' firm.) Soon the Remingtons began receiving orders. Their rolling block wasn't a repeater, but it was a reliable gun that even a minimally trained soldier could handle. Rolling blocks were cost effective, simple, and appealing to most militaries that were still years away from adopting repeaters. Spanish authorities in Cuba wanted 10,000 of them, Japan 3,000, Peru 5,000, and the Papal States ordered 5,000. Eventually scores of thousands of rolling blocks found homes throughout Europe and the Americas, saving the Remingtons from the post–Civil War gun-business malaise.

One customer who developed a fondness for both the rolling block and Samuel Remington was Khedive Ismail Pasha, ruler of Egypt, who aspired to turn northeast Africa into an Egyptian empire, with Cairo serving as the Paris of his imperial domain. Egypt's old muzzleloaders were hardly the kind of weapon a modern nation should have, especially one ruled by an aspiring conqueror, and Ismail was on a quest to replace them. A commission he sent to Europe thought the rolling block looked like a good candidate, so he invited Samuel Remington to Cairo for tests. Everything there went so smoothly that in June 1869 the khedive ordered sixty thousand of them. Then he gave his new friend Samuel Remington land in Cairo's chic residential district where he could build a mansion, which he did, because to have not done so would have been an insult. Remington's social ways did not wither in Cairo. His home became a locus for the American colony's winter social scene, where he provided good times to all, including former Confederates, because, as one old soldier observed, "there is no North or South here."

With the Remington company on the upswing again in 1870, brother Philo, presumably a man of simpler tastes, decided it was time to build a palace of his own in upstate New York. He constructed an Italianate three-story mansion with a tower suitable for a castle, as well as a fountain, and stables as opulent as the house. He erected it on aptly named Armory Hill, overlooking the town of Ilion and its ever-growing factory, the heart of the Remington empire.

Orders for rolling blocks kept coming in. The year Philo raised his mansion, France launched a war against Prussia, and since the French found themselves not armed well enough for such a venture, they had to find guns elsewhere. The Remingtons were ready. They diverted to France a few thousand rolling blocks intended for Egypt and then served as agent for Winchester, which sold the French 4,406 of its Model '66 lever-action repeaters, as well as Spencer rifles and carbines the Winchester company had acquired when it bought the remnants of that firm just the year before. Had the Cheneys held on a little longer, perhaps they, rather than Winchester, could have sold their company's stored Spencer rifles to the French, reaping windfall cash before the firm finally went under.

Between September 21, 1870, and May 6, 1871, the Remington armory in Ilion sent a total of 214,247 guns to France. To get this done, the machinery on its four acres of factory ground devoted to making small arms had been manned by more than a thousand workers for twenty hours of each business day. It would take four years for British companies to furnish their army with three hundred thousand breech-loading rifles of a new British design, according to published estimates, leading the United States *Army and Navy Journal* to crow that it was "not surprised that the London *Times* should seek to arouse English public sentiment by the statement that one American small-arms manufactory has a larger productive capacity than those of all England combined."

The bout over who would control the New Haven Arms Company was tame compared with the battle between Smith & Wesson and Rollin White. Like Henry, White was smarting from the deal he had struck with people who he thought were unfairly enriching themselves with his invention. Money poured into the bank accounts of Horace Smith and Daniel Baird Wesson, while White had to spend his own cash defending his bored-through cylinder patent against infringers. He thought he deserved more—a lot more—and he pressed his case as far as he could take it. Before White's campaign for a bigger piece of the pie ended, he and his cylinder patent would be scrutinized by the chief of ordnance, both houses of Congress, the Supreme Court, and the president of the United States.

The White patent that Smith & Wesson used to monopolize the cartridge revolver market was set to expire in 1869, an event awaited by other manufacturers eager to make guns with bored-through cylinders without fear of being sued. There was a chance the patent could be extended, as Samuel Colt's revolver patent had been, but bad blood between White and Smith & Wesson made planning for that problematic. When the Civil War demand for their guns was greater than the company could produce, the partners wanted to sublicense the patent to other firms. For that they needed White's permission, which he refused to give unless Smith & Wesson doubled the royalties. At one point a possible deal to allow Colt to make its revolvers with bored-through cylinders was scotched after White insisted on an extra $500,000 fee.

A small but powerful weapon in White's arsenal was another gun he had patented that neither Smith nor Wesson had paid much attention to. It, too, had a cartridge revolver cylinder, though this one loaded from the front and was not bored through. Nobody, except perhaps White, thought his revolvers were safe firearms, much less practical ones, as a witness at the *White v. Allen* trial had recognized when he said, "I should rather be the man to be shot at, than the one to use the pistol." The key to success lay in only one aspect of White's contraption, which Daniel Wesson had realized from the very beginning: the patent for the bored-through cylinder. What Wesson had not realized was that in 1863 a lawyer for White convinced the Patent Office to reissue the front-loading gun's patent with a few changes, including a bored-through cylinder. This meant that anyone making a revolver with that feature had to seek a license from both Smith & Wesson *and* White. What's more, Smith & Wesson needed to get White's approval for the second patent (No. 12,649) before they made revolvers with bored-through cylinders, and White couldn't produce guns like that—which he was doing at his own new gun company—unless Smith & Wesson allowed him to use the first (No. 12,648). Two patents for the same thing was an odd situation, to say the least. What White wanted out of it apparently was leverage. What he got was a fight, but not immediately; for a while, both White and Smith & Wesson let the matter lie and continued business as normal.

On September 10, 1866, two and a half years before both cylinder patents were due to expire, White applied for an extension that would give him a few more years of patent protection. But he applied for only one patent—No. 12,649—not the one Smith & Wesson had been using. If the partners had not eventually discovered that their precious patent was not in the pipeline to be renewed, theirs would have expired and White would have had exclusive right to make revolvers with bored-through cylinders. But they found out and pushed White to apply for an extension of No. 12,648, which he did. By then the US Supreme Court had finally decided Ethan Allen's appeal of Judge Clifford's decision that White's first patent was valid, but it did so with four justices agreeing with Clifford—including Clifford himself, of course—and four justices favoring Ethan Allen. When a Supreme Court vote is evenly split, the decision at the trial level stands.

Then in late March 1869, just days before White's patent was set to expire, the commissioner of patents refused to extend it. White asked Congress to step in, claiming that he had been unfairly deprived of profits from his invention because he had to spend his royalty money on lawsuits. Without debate the House and Senate quickly passed a "Bill for the Relief of Rollin White," calling on the Patent Office to give him another hearing. Unfortunately for White, Chief of Ordnance Alexander B. Dyer was against the patent renewal and said so in a letter to the secretary of war, which he hoped would find its way to the desk of President Ulysses S. Grant. Despite White's poor-mouthing, Dyer wrote, he had already received nearly $70,000 in royalties. Besides, "the government suffered inconvenience and embarrassment enough during the war in consequence of the inability of manufacturers to use this patent."

In other words the legal actions White was obligated to take against infringers had hurt the Union in its fight to conquer the Confederacy. On January 11, 1870, Grant vetoed the bill, citing Dyer's letter. White's attempt to have Congress overturn the presidential veto failed. Now that White's patent was dead, a host of companies started producing cheap cartridge revolvers without fearing they'd be sued for infringement.

For most people that probably would have been the end of it. But White was nothing if not tenacious. He returned to Congress, where he pushed for another bill to get a patent rehearing and was making headway until he faced serious opposition from a group calling itself the "Remonstrants," which filed a brief with the House Committee on Patents urging that the bill be killed. The brief didn't identify who the Remonstrants were, but the lawyer filing it had represented Smith & Wesson, and in it was an addendum by Daniel Wesson himself attacking White's claim that he had not been adequately rewarded. The Remonstrants didn't limit themselves to challenging White's patent. With stunning vitriol, they went after the man himself. White, they said, "seems to have a constitutional aversion to doing anything in a straight forward or honest manner. We presume this is natural with White, and that he is to be excused on the same principle that the mother excused her lazy boy, that he was 'born tired,' and could not help it!"

As far as the Remonstrants were concerned, White's patents were as worthless as he was. "For five years have they been dead, and by this time they 'stinketh.' For five successive years have these rotten corpses of what were but abortions of inventions been paraded before Congress, compelling the manufacturers to dance attendance here, spending time and money, to prevent the resurrection of these patents, which it is clear he never was entitled to." To make their case, the Remonstrants referred to earlier examples of bored-through cylinders that they said should have barred White from receiving a patent. By backing that position, Wesson was saying that the exclusive patent that he had used to enrich himself and Horace Smith had been invalid from the very beginning, just as Ethan Allen had argued in the court battle they had waged against him. Even worse, the Remonstrants claimed, White had tried to bribe the patent examiner. "[W]e are prepared to prove," they wrote, "that on the day of the hearing one of his friends stated that *a check for fifteen thousand dollars was placed in his hands for the purpose of securing the extension!*" This second round in White's legislative crusade for a new patent hearing died in the Senate.

Wesson opposed White's getting the patent extended, because he no longer needed it, and he certainly didn't enjoy the hassles of dealing with a troublesome inventor. Smith & Wesson was on to bigger things, literally bigger. For years the company had been exploring the possibility of making a more powerful cartridge revolver, something it could sell to the military, and while White was lobbying for an extension, the firm had the perfect candidate, safe from competition after patents had been bought from men who had invented useful components. It was a graceful, forward-jutting revolver whose barrel and cylinder pivoted down when a latch was released, popping out six spent cartridge casings all at once and allowing fresh loads to be inserted before the shooter rolled the barrel-and-cylinder unit back into place. This was faster than the one-at-a-time ejection system of Colt and Remington cartridge revolvers. The Ordnance Department had tested it and ordered a thousand, making the Smith & Wesson No. 3 the first revolver in service to use metal cartridges. Even better for the company, Russian Grand Duke Alexis Alexandrovich liked it. On a visit to the United States, he stopped by the Smith & Wesson factory, where the company delighted him with a gift of an engraved No. 3 with pearl grips in a presentation case. He had it with him when hunting buffalo with Buffalo Bill Cody and George Armstrong Custer. The Russian government also liked it enough to place an initial order for twenty thousand of them. There would be more later.

As they did with most new firearms, Smith & Wesson made continual improvements to the No. 3 during production. Suggestions came from a number of sources, including the Russian military and American soldiers in the field. As Samuel Colt had Texas Ranger Samuel Walker to give him ideas for fixing the Paterson revolver, Smith & Wesson had Maj. George W. Schofield to offer his thoughts on improving the No. 3. Major Schofield took an interest in the new Smith & Wesson from the first time he heard of it, perhaps from his illustrious older brother, Gen. John M. Schofield, president of an Army Small Arms Board when the No. 3 was in its infancy. George had ambition as well as a gift for mechanics, so in 1870 he approached Smith & Wesson, offering his services as an agent for the company in Colorado and western Kansas. The major must have seemed an important future asset for the company, because the partners sent him one of the first hundred-count batches of No. 3 revolvers completed at their Springfield factory, along with five hundred cartridges, all free of charge. They couldn't give him an exclusive agency, they said, but they would sell him as many handguns as he wanted at a much-reduced price. Schofield understood what a cavalryman needed in a gun, and while he thought the No. 3 was a good one, he knew he could make it better. First he patented a barrel latch that made the revolver easier to reload on horseback. Next he worked on improving the part that extracted spent cases, winning himself another patent. After Smith & Wesson incorporated his modifications, Schofield tested the guns and praised them to his military colleagues.

In 1873 Daniel Baird Wesson became the sole owner of Smith & Wesson, Horace Smith having sold him his interest in the firm and retired. Also that year Daniel Wesson found himself once again squaring off against Colt. The deceased Samuel Colt wasn't his opponent, of course, but his company was, with a new cartridge revolver of its own that it wanted the Army to adopt. In July 1873 the Ordnance Department picked the Colt, partly because it was deemed sturdier, though Smith & Wesson was able to convince the department the next year to buy three thousand of the revolvers now called the Schofield. The military wanted Schofields made to fit the .45-caliber cartridge used in the Colt revolver, something Daniel Wesson said was impossible because of his gun's case extractor system. He might have redesigned the Schofield to accommodate the Colt round, but he focused instead on making a shorter cartridge that would work in both guns. This would prove to be a problem.

Immediately after returning from his second expedition against Indians in Oklahoma Territory, George Schofield wrote to Wesson from Fort Sill to report on how his pistols performed. "The one I have (altered from your Russian model) has stood every test. Have killed almost everything—but an Indian—with it and used it in all kinds of weather and in the roughest sort of a way. On the last trip I loaded twice with my horse at a run and killed a buffalo each time."

It looked like the Army would have two revolvers in regular service on the frontier, one Colt and the other the Smith & Wesson Schofield, until some mix-ups in ammunition deliveries revealed Daniel Wesson's mistake. Smith & Wesson cartridges could be used in Colt revolvers, but the bigger Colt cartridges couldn't fit in the Schofields. If the wrong shipment of cartridges arrived at a government outpost, it wouldn't matter if the soldiers had Colts, but if they were armed with Smith & Wessons, they would be without usable ammunition. The Colt was favored, but both guns stayed in service, with the Smith & Wesson No. 3 continuing to sell well, including in other countries, such as Japan, Argentina, and Turkey.

George Schofield kept at least one of his namesake revolvers for himself. He also kept alive his interest in mechanical things, a trait some fellow officers frowned on, because they thought it conflicted with his duties. Nevertheless, Schofield eventually became a lieutenant colonel in command of the Sixth Cavalry at Fort Apache, Arizona Territory, a lonely outpost made more so by the death of Alma, his twenty-four-year-old wife of only four years, in late March of 1879 while they were at Fort Sill. Shortly thereafter he began to have health problems of his own. Late in the year his wife died, Schofield injured his knee when his horse fell, and he was kept on sick leave until well into 1880. For several days in December 1882, Schofield seemed nervous. In one conversation he was cheerful enough but sounded a bit irrational. At daybreak on Sunday, December 17, he put on his dress uniform, combed his hair at his washstand, pointed the muzzle of his Schofield revolver at his right eye, and sent a bullet through his head. The Fort Apache surgeon concluded that "a fit of temporary mental aberration . . . probably brought on by fatigue and worry, and the intense pre-occupation of his inventive genius" drove the forty-nine-year-old Schofield to suicide.

30

CENTENNIAL

[It] is still in these things of iron and steel that the national genius most freely speaks; by and by the inspired marbles, the breathing canvases, the great literature; for the present America is voluble in the strong metals and their infinite uses.

WILLIAM DEAN HOWELLS. "A SENNIGHT OF THE CENTENNIAL," *THE ATLANTIC MONTHLY*, JULY 1876

With the pull of two levers, the largest stationary steam engine in the country hissed to life. Concrete beneath the giant beast began to vibrate. Then its huge walking beams started their steady journeys up and down, round and round, stirring hundreds of slumbering smaller machines, feeding them power to begin their work, some printing newspapers or wallpaper, others sewing or making twenty-five thousand shingles in a single day. The monster machine was a Corliss steam engine, symbol of the rising industrial might of the civilized world. It was also a new symbol of the United States, a country celebrating its first century of existence, while at the same time announcing to the world that it was ready to take charge.

The Corliss took center stage in the thirteen-acre Machinery Hall at the vast Centennial Exhibition in Philadelphia, dominating everything around it from the opening ceremony in May 1876 all the way to the Centennial's closing the following November. Not only was the Corliss imposingly massive—45 feet high and weighing about 680 tons—but its improved operating system was making steam power triumph over waterpower, especially in the Northeast, freeing factories from dependence on streams and rivers, allowing their owners to build wherever they wished. Manufacturing, long tied to Eastern waterways like the Connecticut River and its tributaries, could move westward more quickly.

"There it stands," exclaimed one chronicler of the moment, "holding its place as a veritable king among machinery, so powerful and yet so gentle, capable of producing the most ponderous blows upon the anvil, or of weaving the most delicate fabrics; that to which all other machines must be subservient, and without whose labor our efforts would be small indeed; the breathing pulse, the soul of the machinery exhibition." Americans—those accustomed to large engines such as locomotives, as well as those familiar with nothing bigger than a corn sheller—liked bigness in mechanics, as they had shown a generation earlier when rapturously gazing at Springfield Armory's gun-making machinery. The giant Corliss, with more than a mile of piping that brought energy to all the chattering, humming machines within its reach, suited the zeitgeist. It was said that when poet Walt Whitman toured the Centennial, he sat for a half hour before the Corliss in awestruck silence.

The Centennial event was the first official world's fair to be held in the United States, and the country wanted to make the most of it. Reconstruction was under way with wounds from the Civil War still in need of healing more than a decade after the South had surrendered. The Panic of 1873, yet another violent economic downturn, which started in Europe and spread quickly to the United States, brought industrial expansion to a halt, sending the economy reeling for years. Adding to the nation's woes were widespread corruption scandals from President Grant's administration on down. The Centennial celebration gave the country a chance to restore its spirit as well as tout its achievements.

For the occasion a veritable city of exhibitions rose on the site, with buildings showcasing all manner of modern wonders, from locomotives to the delicate mechanisms of watches. In all there were over thirty thousand exhibits, far more than the seventeen thousand at the 1851 Crystal Palace exhibition in London. Next to a motor that held promise for running sewing machines stood "a most interesting exhibit of asbestos, a mineral which has the peculiar property of being a non-conductor of heat." Near a restaurant where one could dine well for fifty cents, "the pop-corn man had a tasteful stand, from which he does a thriving business in this peculiarly American eatable." Elsewhere at the exhibition, two single-story frame buildings were devoted solely to popcorn that concessionaire J. A. Baker hoped would help justify the $7,000 fee he had paid for the "Exclusive Right to Sell Pop-corn." (The right to sell soda water went for $20,000.) *The New York Daily Herald* labeled Baker "a popcorn capitalist" willing to spend exorbitantly "for the sole privilege of impairing the digestion of the world at the great fair."

In a small factory with everything needed to make chewing tobacco except for extra flavoring, four black men twisted and pressed the leaves while fairgoers watched. "The negroes, as they work, sing the songs and hymns which are familiar to those who have visited the tobacco factories of the South," recounted a contemporary history of the exhibition. The colossal forearm and torch-bearing hand of the future Statue of Liberty held visitors seeking a loftier view of the exhibition. (*The New York Times* jocularly noted that Liberty's hand was "of such enormous proportion that the thumb-nail afforded an easy seat for the largest fat woman now in existence.") A young Thomas Edison showed off a few of his devices, including an "automatic telegraph system" and an "electric pen." A Scottish immigrant named Alexander Graham Bell demonstrated a machine he had invented to transmit voices over electric wires. When he heard human speech emanating from Bell's telephone, Centennial visitor and Brazilian emperor Dom Pedro II raised his head from the receiver and blurted, "My God, it talks!"

And, of course, there were weapons. "Almost every country showed arms or processes for making arms," one scholar wrote. "The displays ranged from old-fashioned matchlocks, chain mail, and daggers used by natives of

India to the latest monster cannon forged in Germany." An Englishman living in New York told his brother back home that he encountered "all sorts of Guns of all sizes up to the 20-in. 'Dahlgren' with a heap of its 1080 lb shot" arrayed outside the American government building. Inside "there is every conceivable instrument for scientific murder, fire arms, old and new, shot & shell, whole & sawed in two to show the interior, edged weapons, pikes, torpedoes. . . ." Primitive implements of war at the exhibition filled an eighty-four-page report titled "A study of the savage weapons at the Centennial exhibition, Philadelphia, 1876."

In the government building the Ordnance Department set up arsenal machinery, where skilled workmen made live, assembly-line, high-powered rifle ammunition for the entertainment of spectators. Visitors could pick up little souvenir boxes containing pieces that showed the stages of a cartridge-in-the-making, including cases and bullets. At the Springfield Armory exhibit, fairgoers watched rifles being made. There "skilful men operatives begin with the round bars of steel and the long blocks of black walnut, turning out complete the handsome weapons of death almost as rapidly as the latter could be made to end human lives," claimed an exhibition observer. Springfield Armory workers also made ammunition and sharpened bayonets on fine-grained grinding wheels that sent forth streams of sparks. Cases filled with "guns and bayonets of all patterns" covered the walls at the rear of the government building. "There were pistols and revolvers enough to arm the Russian soldiery, and of so many different, odd and pretty styles that all the tastes on earth could make gratifying selections."

As expected, the big private gun companies showed up. More than three hundred revolvers—most of them engraved, some nickel-plated or with gold parts and pearl grips—stood in concentric circles on Colt's ornamental case. In two separate buildings Colt presented its prized fast-firing, multi-barreled Gatling guns, considered a predecessor to the machine gun and already adopted by Russia, Turkey, Egypt, and some American military units. Centennial judges bestowed a medal on its inventor, Richard J. Gatling. Smith & Wesson sent an array of elaborate pistols, among them a richly gold-inlayed, pearl-handled revolver that was to become a well-traveled veteran of exhibitions. For its part the Winchester company displayed nearly two hundred guns that included a rifle in an exhibition case, its blued steel inlaid with gold tracery, with a mirror to let visitors see the entire thing from all sides. The company also introduced a hefty lever action rifle called the "Centennial" that could hold its own with the increasingly powerful long arms of the day. Exhibition judges gave Winchester two medals. One was for "the best magazine rifle for sporting purposes yet produced," the other for perfection in military ammunition.

E. Remington & Sons, another two-medal winner, brought guns that *Frank Leslie's Illustrated News* said were "as handsome specimens of work and finish as any to be found in the Exhibition." Between groups of gun cases there stood a large star made from Remington cartridges. This was flanked by radiating displays of edged weapons strategically placed over samples of sixteen different Remington pistols. Above the cases rose a three-foot-high signboard with the firm name spelled in letters made from large, nickel-plated revolvers that "not only make up very elegantly in form, but have a very pleasing effect against the purple velvet with which the sign as well as all the showcases and panels are covered." Atop everything the Remingtons had mounted an American eagle surrounded by flags of those governments that had used or adopted their rolling block rifle. The names of the governments were laid at the eagle's feet.

Despite the economic crisis brought on by the Panic of 1873, these major gun firms were picking up speed, and the nation would soon follow. Some of the firms were expanding their product lines. At the Centennial, for example, the Remingtons showed off a "typographic" machine (typewriter) that carried the company name. For fifty cents a visitor could buy a typewritten letter to send to a friend in another town, a good advertising ploy. The firm was also making sewing machines. The men in charge of these companies had placed themselves at the forefront of business development and innovation, while enhancing their personal wealth far beyond what was possible for the cottage gunsmiths or armory mechanics of prior generations. For Colt, Winchester, Remington, and Smith & Wesson—for the country itself—it was progress on nineteenth-century European-American terms.

Their captains had become masters of industry, charter members of a dawning Gilded Age.

Those who first came to the continent, the Indians, were not neglected at the Centennial, though they weren't treated as first-class citizens. The year before the exhibition opened, the Smithsonian Institution's assistant secretary, Spencer F. Baird, wanted an exhibit to "illustrate the past and present condition of the native tribes of the United States, or its anthropology." Baird had two motives. One was to enlighten the public about Native Americans. The other was to use the Centennial to expand the Smithsonian's ethnological and archeological possessions. There was even talk that the Institution would host thirty Indian families on a five-acre reservation adjoining the Centennial grounds. "The presence of a band of uncivilized aborigines of America at the Centennial will form an interesting feature," observed several US newspapers. Congress was approached about creating a temporary Indian village in Philadelphia, but no funding came.

There were mannequins, though, including one that fairgoers thought was of Sioux chief Red Cloud. Not surprisingly, to some visitors he was an ominous presence, dressed in warrior attire with a belt of human scalps, a tomahawk in his raised hand. One newspaper said he seemed "ready to pounce on some unsuspecting victim." Smithsonian officials had hoped their Indian displays, which turned out to be quite popular, would nurture understanding, despite the racism of the times and the crowded presentation without context. They viewed collecting artifacts at least partly as a means to preserve remnants of a culture that modernity was extinguishing. Even people who counted themselves as friends of the Indian believed the native way of life belonged to an era that was of necessity vanishing, a primitivism that had to give way to the kind of progress in which the Centennial exulted.

Walking past the Exhibition's Agricultural Hall one mild July afternoon, the assistant to the man in charge of the Smithsonian's ethnographic display stopped to look down at a piece of quartz sticking aboveground next to his path. Frank Hamilton Cushing had only just turned nineteen, but he possessed a keen eye for old things, and this rock caught it. Cushing's scholarly love of American Indian objects and people was already forming and would eventually take him to the Southwest, where he would become a devoted member of the Zuni tribe, about whom he would learn and write. It could be said of him later that he had "gone native." At this moment, however, all his attention was on that bit of quartz.

The nugget had probably been plowed up weeks earlier, when the Centennial Exhibition was being built, and had lain unnoticed by visitors as they walked by. The stone would not attract most people's attention, but to Cushing, its rough surface had the look of something fashioned by human hands. He picked it up, and when he had brushed away surface grime, he saw that it was the point of an Indian club, its shape formed by a stone hammer that had chipped away at it at least five hundred years earlier.

"It is certainly curious," *The Philadelphia Inquirer* observed, when it learned of Cushing's discovery, "that in the midst of all the wonders of ancient and modern times, and in proximity to some of the grandest triumphs of mechanical art, this implement, manufactured centuries ago by the original inhabitants of this country, should suddenly reappear, the reminding of another age and another people, provides the strongest imaginable contrast to the mechanical progress of the present."

31

SACRED IRON

The Centennial Exhibition had been under way for a week when Lt. Col. George Armstrong Custer and his Seventh Cavalry left Fort Abraham Lincoln in Dakota Territory, heading west into Montana. The Grant administration's goal was to force the Plains Indians out of the way of progress, and the Seventh Cavalry was to encourage them to go.

Custer was no Indian hater. Instead his personal quest, as always, was for success in combat—and now for redemption after his recent embarrassment of being told he could not go on the 1876 campaign for several reasons, including his testimony about corruption within the Grant administration. Roasted by the press and told by top military brass that Custer was needed for the Indian campaign, Grant relented and allowed him to go as subordinate to General Alfred Terry. Now Custer and two other commands were on their way.

Complicating matters—at their core, really—was gold. In 1874 Custer and the Seventh Cavalry led an expedition to the Black Hills, looking for a suitable place to build a fort and to see what natural resources were there. They found gold. When word of the discovery spread, gold fever lured prospectors, hordes of them, to invade the Black Hills, land the United States had guaranteed to the Lakota and Northern Cheyenne just six years earlier. A windfall of mineral wealth could ease the economic pain the country was feeling in the wake of the Panic of 1873.

The federal government had tried to buy the Black Hills and send the Indians elsewhere, but negotiations fell apart. In the end the Indians wanted to stay. "You speak of another country, but it does not concern me," Brulé Sioux chief Spotted Tail told the government. "I want nothing to do with it. I was not from there; but, if it is such a good country, you ought to send the white men now in our country there and let us alone."

There had been skirmishes before, even serious battles, as encroachment by settlers, buffalo hunters, and the railroad continued relentlessly. The rush for the Black Hills escalated everything. The Great Sioux War was on.

Plains Indians had long been gathering weapons, symbols of manhood in a culture where warfare and hunting were central. When the Civil War ended, they took advantage of a new flood of guns coming their way. Repeating rifles joined cheap military surplus arms in a firepower bonanza for Indians, who fought each other as well as the swelling population of light-skinned settlers arriving from the East. Traders, some licensed, many operating illegally, were happy to make a few dollars selling guns to the tribes. A horse or a mule could pay for a repeating rifle; buffalo hides were good to trade for precious ammunition, which was usually in short supply. The US government did its part by auctioning off small arms it no longer needed. Indians often received guns as goodwill gifts for hunting as part of treaty obligations. Then there were the rifles sold legitimately to "friendly" Indians for hunting but which eventually made their way into the hands of "hostiles" less disposed to living peacefully with whites. "The Sioux must have good white men friends on the Platte and Missouri," reported a Crow chieftain in 1873 after a skirmish with the Crows' longstanding Indian enemy. "They get guns and ammunition; they are better armed than we are; they have Winchester, Henry, and Spencer rifles and needle guns. We took some of these guns from those we killed; we took two Henry rifles and one needle-gun."

The natives' enhanced firepower hadn't escaped Custer's notice that year. "The arms with which they fought us (several of which were captured in the fight) were of the latest improved patterns of breech-loading repeating rifles, and their supply of metallic rifle-cartridges seemed unlimited, as they were anything but sparing in their use," he wrote in his report on engagements with Indians beside the Yellowstone River in Montana Territory. "So amply have they been supplied with breech-loading rifles and ammunition that neither bows nor arrows were employed against us."

Plains warfare had been undergoing a sea change, first when firearms in general showed up, giving some tribes an edge over others and raising the stakes in intertribal arms races. By 1876 about half the Lakota warriors had some kind of firearm, including obsolete flintlocks. Of all the guns in Indian hands then, a fifth were modern repeaters, which elevated the warriors' ability to kill. When Indians began using revolvers to fight the white man, combat became more equal than it was when Jack Hays had fought Comanches thirty years earlier. Now, with Winchesters and Spencers in their arsenal, Native Americans were even more deadly, for it was the breech-loading rifle and metallic cartridges that transformed "the Plains Indian from an insignificant, scarcely dangerous adversary into as magnificent a soldier as the world can show," wrote Army veteran Col. Richard Irving Dodge. "Already a perfect horseman, and accustomed all his life to the use of arms on horseback, all he needed was an accurate weapon, which could be easily and rapidly loaded while at full speed." Now he had one.

The troopers who rode with Custer wore Colt six-shooters on their hips, a far more advanced weapon than the long-obsolete, difficult to reload five-shot Paterson revolvers Jack Hays and the Texas Rangers had used against Comanches. For targets beyond the reach of their short-range revolvers, Custer's men carried single-shot breech-loading carbines called trapdoor Springfields, made at the Springfield Armory. The top of each chamber flipped up on a hinge to allow a fresh cartridge to be loaded. Spencer repeaters had been phased out of regular service, though some of the Army's Indian scouts continued to use them. At Custer's disposal were a couple of Gatling guns, but he decided they would be more hindrance than help on what promised to be fast riding over uneven ground when quick decisions about troop placement might be needed. Mounted on wheels, these heavy, hand-cranked guns were difficult to maneuver. Custer left them behind.

What he did not leave behind was the rolling block sporting rifle he had ordered from the Remington company, a powerful .50-caliber gun that had gone with him on both an 1873 surveying trip in the Dakota and Montana Territories and the 1874 Black Hills expedition to take pleasure in one of his favorite pastimes, hunting. Custer was a good shot and had even chased a buffalo bull at full speed, yelling excitedly, his English greyhounds pacing him, his revolver in his hand, only to mistakenly shoot his horse in the head when the buffalo suddenly veered into them.

Among Custer's personal guns was a Springfield trapdoor rifle built for hunting, which he used on the expedition with Russian Grand Duke Alexis, but now he was partial to the rolling block. In a fan letter to the Remingtons, Custer bragged of "but a portion of the game killed by me: Antelope, 41; buffalo, 4; elk, 4; blacktail deer, 4; American deer, 3; white wolf, 2; geese, prairie chickens, and other feathered game in large numbers." What was remarkable, Custer told the Remingtons, was not that he had killed so many animals with his rolling block, but that he could take them down from hundreds of yards away. "I am more than ever impressed with the many superior qualities possessed by the system of arms manufactured by your firm," he wrote, "and I believe I am safe in asserting that to a great extent this opinion is largely shared in by the members of the Yellowstone Expedition who had opportunities to make practical tests of the question." Despite the hard demands of going after Indians in Montana Territory on this new assignment, Custer thought he might have a chance to hunt again, so the rolling block was with him. He was confident in his mission and didn't think the Indians ahead of him would be unconquerable foes.

"Sitting Bull, thy doom is fast approaching!" exclaimed a correspondent for *The Helena Weekly Herald*, writing from the Yellowstone Valley on June 14. Long Horse, as he called himself in print, had been with the Army for the previous ten days, scouting the countryside for Indians. He wrote that General Terry and Custer were moving up beside the Yellowstone River with twelve companies of cavalry ("the flower of our army"), artillery, and

Indian and white scouts "and will probably strike Sitting Bull in two days." Elsewhere another command of infantry, cavalry, and scouts was on its way. Word had spread that General George Crook was also coming to join in the fighting. "With three military commands, at the heads of which the following well-known names appear—Terry, Gibbon, Custer, Crook, Brisbane, Ball, and others—we cannot see how you can escape," Long Horse warned Sitting Bull. "Better throw up thy hands, old boy, if thou wouldst see the close of this glorious Centennial year."

Custer and twelve companies of the Seventh Cavalry—about seven hundred men in all, scouts included—were far ahead of General Terry when they neared the Valley of the Little Bighorn River on June 24. Custer, now called Hard Ass by some, had pushed the Seventh into riding through the night. By the next morning his scouts were looking down from a promontory onto the prairie that straddled the winding river. In the distance dust clouds and a pony herd, a very large one, told them that an immense Indian village lay ahead, bigger than any they had seen before.

Custer thought his troops had been spotted, so he decided to attack before other commands showed up. He reasoned that if he could drive the noncombatants—women, children, and old men—out of the village and take control of them, the warriors would be more readily managed. Then all could be sent on their way. To accomplish this, he ordered Maj. Marcus Reno and his 140 men to attack from down the river while he approached the village from Reno's right. Captain Frederick Benteen would be behind with another 255 men and a pack train hauling additional ammunition. Around midday Reno charged.

Almost immediately the plan fell apart. Instead of pushing Indians out, Reno halted the charge before he got to the village. Seeing the soldiers stop well short of the village, angry warriors charged back at him, more than a handful armed with repeating rifles. Soldiers scampered in disarray, Reno having lost control and giving conflicting commands. Oglala warrior Eagle Elk, his face painted for war, riding his fastest pony, raced in with his nearly new Winchester, shooting soldiers off their saddles as easily as if he were killing buffalo cows. He shot dead two or three more as they were trying to cross the river. Back at the campground the charismatic Crazy Horse had emerged from his tepee, bridle and Winchester in hand, ready to join the battle.

The troopers' trapdoor Springfields sent bullets farther and harder than most Indian guns did, including the repeaters, but the warriors skillfully maneuvered themselves in close, popping up to shoot, then ducking back down again. Gunfire was so intense, a Crow scout said, that it sounded like "the snapping of the threads in the tearing of a blanket"—and that was just from the variety of guns the Indians used. Warriors rained countless arrows upon the troopers wherever they were. Trapdoor carbines and ammunition captured when Indians overwhelmed soldiers in one part of the battle were turned on the Seventh Cavalry. Benteen's soldiers and the pack train never got to Custer in time to save him and instead joined the survivors of Reno's command on another hill, where they managed to hold on through the night and into the next day until the Sioux and Cheyenne went away.

When Terry's column arrived at the Little Bighorn on June 27, it rescued the remnants of Reno's and Benteen's commands, not by fighting but by its presence, the Indians having moved on after becoming aware of its approach. Around 9:00 in the morning, Terry's advance guard found the dead troopers in Custer's companies, including the former boy general himself, who had been shot twice, once below his heart and once in his left temple.

There was lots of blame to go around for Custer's defeat, much of it directed at Custer himself. The debates continue to this day. Some said that advanced weaponry in Indian hands doomed the Seventh Cavalry. "We advise the Winchester Arms Company," *The Army and Navy Journal* sneered sarcastically a month after battle, to "prosecute the Indians for infringement of their patent." Winchester rifles were "plenty among them; the agency people and the traders solemnly affirm that they don't furnish them; so it can only be inferred that the Indians manufacture them themselves. If Gov. Winchester could get out a preliminary injunction, restraining the Indians from the use of his rifle, it might be of signal service to our troops in the next engagement."

But repeating rifles didn't determine the outcome at Little Bighorn, as Colt's first revolvers had at the Battle of Walker's Creek. It was numbers, tactics, and fearless commitment to a way of life that overwhelmed the Seventh Cavalry, though modern weapons on both sides made killing easier, even as the Lakota found mortal combat the traditional, up-close way more honorable. Numbers and technology would soon work against the Indians, however, making Little Bighorn the last great military victory they would have over the white man. More whites kept moving into Indian lands with increasingly powerful arms, such as the four breech-loading Hotchkiss cannons, among the first of their kind, that the reconstituted Seventh Cavalry used to blast into the Lakota in what would become known as the Wounded Knee Massacre more than fourteen years after Little Bighorn.

The decades that followed the United States' first century saw technology and business increase their stake in ruling the continent together, the titans of American gun industry being part of the vanguard. Craftsmen continued to give way to risk-taking capitalists, who consolidated their power in money and the intellectual property of patents they controlled to create firearms empires that sold their products to the world, as well as to citizens back home. In the years leading up to the Civil War and beyond, private arms makers increased their customer base and, in the process, reduced their dependence on government contracts. Innovation moved from federal armories to private concerns whose creations fed development in non-gun manufacturing. Firms like Colt and Smith & Wesson that were highly capitalized and could operate in the open market in the post–Civil War years flourished, while many smaller shops faded away. By the time all the nineteenth-century gun entrepreneurs had died and their descendants had surrendered control of what they left, manufacturing firearms en masse had become even larger corporate undertakings.

The names Colt, Winchester, Remington, and Smith & Wesson endure today as company identifiers, each calling up visions of earlier eras and individual Americans whose old-fashioned pluck and Yankee ingenuity drove them to make their marks for country and what they saw as progress. Yet whatever truth there is in the message those images convey, the enterprises themselves have become creatures of the modern corporate world, their venerable names an allusion to a past that executives hope will persuade customers in the present to rely on company products as American and trustworthy, even when a company is owned by a non-American entity. Names sell. Samuel Colt would approve.

EPILOGUE

For the rest of his life, **Walter Hunt** continued inventing, though his habit of selling patents to pay debts and feed his family kept him from reaping real rewards from his creations. To end long legal battles over the rights to a sewing machine conceived by Hunt and Elias Howe, Isaac Singer offered Hunt $50,000 in 1858 for his original design, the equivalent of more $1.6 million today. Before any payments were made, however, the inventor died of pneumonia in his workshop the next year at the age of sixty-two.

Texas Ranger Captain **Jack Hays** went to California after serving with distinction in the Mexican-American War. A county in Texas was named for him. Hays became active in politics and Indian affairs and made a fortune in real estate and ranching. The Comanches' respect for Hays never waned. When he learned that Hays's wife had given birth to a son, Penateka chief Buffalo Hump and another Comanche leader gave the former Texas Ranger a gold goblet and two gold teaspoons. On the spoons was engraved, "Buffalo Hump" and "BHH," the latter for Buffalo Hump Hays.

A county in Texas was also named for **Samuel H. Walker**. Actually it was renamed for him after local residents became disenchanted with the original namesake, US senator from Mississippi Robert J. Walker, who had introduced a resolution in Congress calling for the annexation of Texas. During the Civil War Robert Walker favored the Union, an odd stance for a Mississippian, even one originally from Pennsylvania, and an unpopular position in Confederate Texas. To remove that stain and still keep the name, the state decided its Walker County would be for Samuel. On May 24, 2021, the Texas governor signed a resolution passed by both houses of the state legislature declaring the Walker Colt the official handgun of Texas.

After the Civil War **John T. Wilder** settled in Chattanooga, where he was elected mayor just six years after helping to conquer the city. He soon became one of the South's leading businessmen. When he heard that former Confederate general Nathan Bedford Forrest, one of his combat opponents in the Civil War, had been arrested for violating parole by helping organize the Ku Klux Klan, Wilder met with him. Convinced that Forrest was only trying to protect Southerners by heading the Klan, Wilder successfully interceded on his behalf with President Grant. Eventually Wilder became an honorary member of the Forrest camp of the United Confederate Veterans. He had been widower for twelve years when he married the daughter of a Confederate veteran, a woman nearly a half century his junior who was also his nurse. After Wilder put her through medical school at the University of Tennessee, she became the first woman to pass that state's medical exam.

For his last patent **Rollin White** returned to fabrics, inventing an improved spinning-spindle and its support. The US government issued the patent on December 16, 1890, fifteen months before White died at the age of seventy-four. His brief obituary in a Vermont newspaper said, "Mr. White claimed the invention of the Smith & Wesson revolver. A bill to give him possession of the invention was vetoed by President Grant."

After his falling out with Oliver Winchester, **B. Tyler Henry** remained in New Haven, running a machine shop where he continued to invent. In 1889 he patented an axle box for vehicles. According to an obituary in a local newspaper, "He was, just before his last illness, about to perfect a valuable trolley appliance." Henry died at his home, a little over a mile from the Winchester Repeating Arms Company factory, on June 8, 1898, at the age of seventy-seven.

Samuel Colt's factory kept turning out weapons into the twenty-first century, though the company went through bankruptcy reorganization more than once. In May 2021 CZG—Česká zbrojovka Group SE, a firearms maker based in the Czech Republic, bought the Colt firm. American soldiers carried firearms bearing the Colt name in every war. Colt's widow, Elizabeth, became prominent in her own right as a philanthropist and civic leader and continued her devotion to her husband's firm and memory, eventually erecting a church in his honor. In the religious building are artistic motifs in marble depicting Colt firearms.

In 2016 the business started by **Eliphalet Remington** celebrated its bicentennial, marking the legend of the founder's first rifle as its beginning and boasting of being then the country's oldest gun company in continuing operation. Like other firearms companies, the Remington firm went through rough economic times, including major reorganizations in the late nineteenth century. Over the decades the Remington company expanded into products beyond firearms. Eventually when people saw "Remington" they thought not only of guns but also of sewing machines, electric razors, and typewriters. By the late 1880s the Remingtons were out of the family business, the company near collapse. It was saved by a syndicate and continued to produce guns. In 2020 the firm filed for bankruptcy—as it had done more than once before. After having been closed for eight months, the Remington firearms factory reopened in May 2021 with plans to make guns again.

Until his retirement in 1873 **Horace Smith**'s time was devoted to Smith & Wesson, though he occasionally served as a Springfield city alderman. He traveled the American West, studied astronomy, helped his church, and became president of the Chicopee National Bank. When Smith died at the age of eighty-four in his Springfield home in early 1893, he had outlived his son, Dexter, by six weeks. In his will Smith left the bulk of his estate to charities, including libraries, hospitals, the Home for Friendless Women and Children, and the Tuskegee College for Colored Youth in Alabama. His gifts were also small and personal, from barrels of apples for a children's organization to a stove he donated to a women's group to forgiving small unsecured loans he had made.

Daniel Baird Wesson also had interests other than firearms. He became president of the Bigelow Wire Works in Iowa, was among the founders of the First National Bank of Springfield, and acquired a large stake in the New York, New Haven & Hartford Railroad. Wesson also built two hospitals in Springfield, setting aside funds in his will to keep them going. He died in August 1906 at the age of eighty-one; his wife, Cynthia, had died the month before. Their great-grandson, Daniel B. Wesson II, was himself an inventor who launched his own company, Dan Wesson Firearms, which was bought by CZ-USA, a subsidiary of CZG—Česká zbrojovka Group SE, in 2005.

Oliver Fisher Winchester kept angling for government contracts. The Royal Canadian Mounted Police was among the very few government organizations to buy his Centennial rifle in bulk. He always wanted big US military contracts but also pushed hard for lots of civilian interest, which turned out to be a boon for the company. When he died in 1880, the name Winchester had become synonymous with the lever-action rifle. Having, like Horace Smith, become interested in the stars, he left money for an observatory at Yale. The university razed Winchester's mansion to make way for a divinity school. In the twentieth century Winchester's wish was granted: Firearms from his factory were finally purchased by the US government for military service in the World Wars. The firm eventually became the U.S. Repeating Arms Company, which was acquired by the Herstal Group, a Belgian-based firm that continues to make Winchester-labeled firearms in several countries. The New Haven plant closed in January 2006.

Having been in poor health for several years, **William Wirt Winchester** died at the age of forty-three from tuberculosis three months after his father's death. His widow, **Sarah L. Winchester**—one of America's wealthiest women, thanks to shirts and guns—left Connecticut and headed west. She bought a farmhouse in San Jose, California, on which she spent eccentrically until she died in 1922. For decades while she lived and ever since, Sarah and her odd, rambling home—called by promoters the Winchester Mystery House—were sources of curiosity. Legend has it that Sarah kept renovating the place in strange ways after a Boston medium instructed her to do so, lest she be forever haunted by those killed with Winchester rifles. She reportedly held séances to placate the spirits. In a 2010 book on the heiress, California historian Mary Jo Ignoffo debunked many of the tales, saying that Sarah was sane and sensible with only a touch of eccentricity, and that some of her architectural oddities resulted from the 1906 San Francisco earthquake or were designed to compensate for her crippling arthritis. Despite her reclusiveness, she gave generously to charities. Nevertheless, the legend lives. In February 2018 a major motion picture loosely based on her later years and titled *Winchester* arrived in movie theaters. It was a ghost story.

Christopher Miner Spencer kept on inventing guns and other mechanical objects. Among his many brainchildren was an automatic screw machine whose rotating cams enabled unskilled operators to make metal parts rapidly. Spencer never lost his boyhood passion for firearms, serving as president and then secretary and treasurer of the Hartford Wesson Rifle Club. In the 1880s he invented a repeating shotgun and rifle, which a new Spencer Arms Company made and marketed. The guns worked, but the company failed, and Spencer lost heavily. Unlike Colt, Winchester, Smith, and Wesson, he never amassed great wealth, though his many inventions contributed to the success of several non-gun firms. Spencer's enthusiasm and talent had always been discovery, not enrichment. After his wife Dora died in 1882, he married a younger woman who had been his wife's nurse and became a father in middle age. "I'd always hoped to leave you money," he told his family late in life, "but the best I can say is that I don't think I am leaving any enemies." Spencer did leave a son who also became a respected inventor, designing pioneering aircraft and piloting a plane high enough in 1929 to break the light airplane altitude record. Spencer died at the age of eighty-eight, but not before soaring in an airplane designed and built by his son. At the end of his life, he was fit, lean, energetic, and strong, like his Revolutionary War grandfather.

ACKNOWLEDGMENTS

Among the many humbling experiences in writing a book such as this is discovering you knew far less about your subject than you thought you did. A gratifying balance is receiving guidance freely given by those who really do know and whose knowledge you could not hope to match even if you spent the rest of your life studying. For that I can thank the many experts on guns and history who helped me avoid mistakes both of fact and interpretation. Alex MacKenzie at the Springfield Armory National Historic Site was a gracious host and patient teacher. Historian Mary Jo Ignoffo enlightened me on many personal Winchester matters. Author and former director of the National Firearms Museum Jim Supica was generous with his time, especially since I continued to inflict myself on him when he was under deadline pressure for his own work. Ranger historian Matt Atkinson improved my Gettysburg chapter, as ranger historian Mike Donahue bettered my understanding of Little Bighorn. (If I could remember the name of the ranger who found the wallet I dropped at Little Bighorn, I'd thank him again here.) Michael Harrison at the Nantucket Historical Association improved my discussion of whaling, and Danny Michael, associate curator of the Cody Firearms Museum at the Buffalo Bill Center of the West, gave me the benefit of his vast knowledge, as well as worthwhile comments on style. Smith & Wesson historian Roy Jinks opened his voluminous files for me, answering even my stupidest questions with quiet tolerance, and shared some delightful meals with me. Like Roy, Winchester and Colt expert Herb Houze proved as fine a dinner companion as he was a historian, letting me pore over cartons of his records at the public library in Cody, Wyoming, where I made enough photocopies to reduce the lifespan of its copier by years. Herb died in the summer of 2019, leaving an enviable legacy of scholarship that more than occasionally revealed a deliciously light wit. He was a gentleman.

Other friends who happen to be historians, including Larry Peskin, Charley Mitchell, Rita Costa-Gomes, and Laura Mason, vetted draft chapters and improved them in the process, as did former Maryland state archivist Ed Papenfuse. Steve Luxenberg, Stephen Hunter, and Daniel Mark Epstein each spent hours going over my plans for this book, giving me valuable direction when I needed it most. Many trusted friends and former colleagues offered advice that enhanced chapters as I wrote. These include Jim Astrachan, Jenny Bowlus, Stephanie Citron, Alan Doelp, David Ettlin, Gary Goldberg, Neil Grauer, Stan Klinefelter, Pat Meisol, Barbara Morrison, Wendy Muhlfelder, Sarah O'Brien, Barry Rascovar, Bill Reynolds, and Barrett Tillman. Jim Rasenberger, who has become the world's leading expert on Samuel Colt, went through my Colt chapters, correcting mistakes along the way. Many chapters are far better than they would have been had expert editors Dan Zak and Ramsey Flynn not applied their talents to them. Collector and author Robert Swartz provided insight into the legal battles involving Colt and Smith & Wesson. Independent researchers, including genealogists, uncovered information that gave *Gun Barons* depth. Thanks for this to Sue Fox, Ralph Elder, Melanie McComb, and the indefatigable, resourceful, and engaging Jen Cote.

Without the resources of several archives, works like this that dig into the past could not exist. The public in general as well as researchers should be grateful, as I am, that such places are staffed by knowledgeable people who are also a delight to work with. High on that list are Mary Robinson, who recently retired as Housel Director of the McCracken Research Library, and her staff at the Buffalo Bill Center of the West; Mel Smith and the staff at the Connecticut State Library; Sierra Dixon and her staff at the Connecticut Historical Society, particularly Amy Martin; Dave Smith at the Manchester Historical Society; Michelle Tom and now-retired Barbara Goodwin at the Windsor Historical Society; Beth Burgess at the Harriet Beecher Stowe Center; Maggie Humberston at the Wood Museum of Springfield History; Wendy Essery and Madeline Bourque Kearin at the Worcester Historical Museum; the Herkimer County Historical Society, which helped with early Remington material and a description of the area; Regan Miner of the Norwich Historical Society, who put me in touch with most helpful Dave Oat and Richard Russ; Kathy Pierce of the Northborough Historical Society; Ann Lawless, now-retired executive director of the American Precision Museum; and Edward Surato of the Whitney Library at the New Haven Museum and Historical Society. They and their colleagues made long days of research even more pleasurable than did discovery of elusive documents.

Many thanks to all those above for the time and effort they spent helping to make this book worth reading, even making it publishable in the first place. For anyone who contributed but whom I failed to mention individually, please forgive me and know that your assistance mattered. Of course no one other than this author is responsible for any errors *Gun Barons* may contain.

I will ever be grateful to Jim Kelly and to my agent, Rick Richter, both of Aevitas Creative Management, for seeing value in what I was doing and presenting it to Charles Spicer at St. Martin's Press, and to both Charles Spicer and Sarah Grill, who helped transform the raw material into a book I hope is worthy of them. Thanks, too, to Kathy Harper for helping me craft the formal proposal and make the finished manuscript ready for submission to the publisher.

The steady support of my children over the years was crucial. Thank you, George, Garrett, Clayton, and Julia for being the shining stars of my existence. Finally, a special thanks to Theresa Peard, who not only enriches every moment of my life but who spent innumerable hours over the past couple of years helping me write, research, and, most importantly, think. Without her inspired devotion, this book would not be in your hands.

NOTES

INTRODUCTION

1 "Gun owning was so common": James Lindgren and Justin L. Heather, "Counting Guns in Early America," *William and Mary Law Review* 43, no. 5 (2002): 1840–41.

1 In the arms-bearing society: Amy Ann Cox, "Depending on Arms: A Study of Gun Ownership, Use, and Culture in Early New England" (Ph.D. diss., University of California, Los Angeles, 2010).

1 And the right to hunt belonged to all: David Harsanyi, *First Freedom: A Ride Through America's Enduring History with the Gun* (New York: Threshold Editions, 2018), 24–25.

2 remained virtually unchallenged: Angela Frye Keaton, "Unholstered and Unquestioned: The Rise of Post-World War II American Gun Cultures" (Ph.D. diss., University of Tennessee, Knoxville, 2006).

2 Some museum curators refused to add: Barbara Eldredge, "Missing the Modern Gun: Object Ethics in Collections of Design" (MA thesis, School of Visual Arts, New York, NY, 2012).

2 By creating what some in the twenty-first century: The term "assault weapon" or "assault rifle" has had various meanings. Sometimes it refers to a firearm that can shoot repeatedly with a single pull of the trigger. Often it is used to describe a semi-automatic gun with cosmetic features that make it look like a military weapon, even though it doesn't function like one.

5 an intellectually restless man: Joseph Nathan Kane, *Necessity's Child: The Story of Walter Hunt, America's Forgotten Inventor* (Jefferson, NC: McFarland, 1997); Jimmy Stamp, "The Inventive Mind of Walter Hunt, Yankee Mechanical Genius: The compulsively creative Hunt might be the greatest inventor you've

never heard of," at smithsonian.com, October 24, 2013. www.smithsonianmag.com/arts-culture/the-inventive-mind-of-walter-hunt-yankee-mechanical-genius-5331323/; and J.L. Kingsley, "Death of a Prominent Inventor," *Scientific American* 1, no. 2 (July 9, 1859): 21.

8 Machinery and technology from there: Bruce K. Tull, "Springfield Armory as industrial policy: Interchangeable parts and the precision corridor" (Ph.D. diss., University of Massachusetts, Amherst, 2001), 2.

8 "The general arrangement of the interior": excerpt from a letter Wade wrote to Col. Roswell Lee, May 13, 1830, quoted in Alan C. Braddock, "Armory Shows: The Spectacular Life of a Building Type to 1913," *American Art* 27, no. 3 (Fall 2013): 44.

9 "The whole scene appears more beautiful": "U.S Armory, at Springfield," *Hampshire* (Northampton, MA) *Gazette*, September 27, 1837.

9 "The machinery here is absolutely poetical, both in structure and operation": "Springfield Trade and Industry. No. 7. United States Armory," *Springfield* (MA) *Republican*, August 5, 1851.

9 more than a hundred thousand new muskets: Springfield Armory Proceedings, Series VI.A.2, Box 01, Folder 26 at the Springfield Armory National Historic Site, Springfield, MA.

1: DEVIL YACK

13 Yellow Wolf knew the Pinta Trail well: Yellow Wolf is not identified in reports at the time, but some later secondary sources give that name for the Comanche chief killed at the Battle of Walker's Creek.

13 From a wooded hill above: There has been much mythmaking about the Battle of Walker's Creek over the decades. And while the fighting was the revolver's first combat trial when everyone on one side had a five-shooter, at least one Comanche had seen a demonstration of it in 1839. See Thomas W. Kavanagh, *The Comanches: A History 1706–1875* (Lincoln: University of Nebraska Press, 1996), 268–69.

14 Texians were familiar with: "Texian" was a term applied to people, particularly Anglos, who lived in independent Texas before it became a state.

15 A woman who helped bathe: The quotes by Mary Maverick come from *Memoirs of Mary A. Maverick*, Rena Maverick Green, ed. (San Antonio, TX: Alamo, 1921), 44.

15 "her head, arms and face": Whether Matilda Lockhart was in the terrible condition Maverick claimed has been debated. Official reports and family correspondence don't mention her appearance. See Stephen Harrigan, *Big Wonderful Thing: A History of Texas* (Austin: University of Texas Press, 2019), 212.

16 On their order soldiers Conflicting reports about the Council House fight

and Matilda Lockhart have been circulating since it took place. For a recent discussion see Cristen Paige Copeland, "What Went Wrong? How Arrogant Ignorance and Cultural Misconceptions Turned Deadly at the San Antonio Courthouse, March 19, 1840" (master's thesis, University of North Texas, Denton, 2008).

17 Captain John Coffee "Jack" Hays: Descriptions of Jack Hays can be found in Samuel C. Reid, Jr., *The Scouting Expeditions of McCulloch's Texas Rangers* (Philadelphia: G. B. Zieber, 1847), 108–9.

18 "no officer ever possessed more completely": John Salmon Ford, *Rip Ford's Texas*, ed. Stephen B. Oates (Austin: University of Texas Press, 1987), 64–65.

18 The Lipan Apaches, who were no friends: For a discussion of intertribal conflict and a critique of white historians' "ethnocentric viewpoint" that the Indian wars of the American West meant only Indian-white conflict, see John C. Ewers, "Intertribal Warfare as the Precursor of Indian-White Warfare on the Northern Great Plains," *Western Historical Quarterly* 6, no. 4 (October 1975): 397–410.

18 "Captain Jack heap brave": Walter Prescott Webb, *The Texas Rangers: A Century of Frontier Defense*, 2nd ed. (Austin: University of Texas Press, 1965), 65. The quote sounds like something an Indian in an ancient Hollywood Western might say, but it's included here, because Webb is an early source for it.

19 "I will not become a party to a negotiation": excerpt from a letter written by Dudley Selden to Samuel Colt, quoted in William B. Edwards, *The Story of Colt's Revolver: The Biography of Col. Samuel Colt* (New York: Castle Books, 1957), 82.

21 "Any man who has a load, kill that chief!": Robert M. Utley, *Lone Star Justice* (New York: Berkley, 2002), 12.

22 The Paterson revolvers "did good execution": Hays's report on the battle comes from the *Journals of the Ninth Congress of the Republic of Texas, Appendix* (Washington, DC: Miller & Cushney, 1845), 32–33, Texas State Library, Texas Congress House *Journals*, Box no. 1791.6, J826 9th.

2: RISE OF THE SHOWMAN

23 "To all whom it may concern": excerpt from Colt's patent petition for a submarine battery, Samuel Colt Papers, Box 6, Connecticut Historical Society (CHS), Hartford, CT.

23 "*Sam'l Colt Will Blow a Raft Sky-High*": William B. Edwards, *The Story of Colt's Revolver: The Biography of Col. Samuel Colt* (New York: Castle Books, 1957), 18.

24 And dazzle them he would: A recent and comprehensive biography of Colt is Jim Rasenberger's *Revolver: Sam Colt and the Six-Shooter That Changed America* (New York: Scribner, 2020).

24 While at Amherst Academy: The account of Colt shooting off a cannon at Amherst Academy comes from Herbert G. Houze, *Samuel Colt: Arms, Art, and Invention* (New Haven, CT: Yale University Press, 2006), 37, quoting a letter from Edward Dickinson (Emily's father) to Henry Barnard, July 22, 1864, Samuel Colt Correspondence, Wadsworth Atheneum Museum of Art, Hartford, CT.

25 It was during this voyage: Accounts of how Samuel Colt came up with his revolver idea are many, some put out by the inventor himself, such as the one that has him inspired by the turning of a ship's wheel or perhaps its windlass. He also could have seen a flintlock pistol with a many-chambered cylinder that had been patented by Boston-born but London-based Elisha Collier in 1818.

25 "The public are respectfully informed": *Baltimore Patriot & Mercantile Advertiser* 35, no. 109 (May 7, 1830): 3.

25 The advertisement about the gas's "most astonishing effects": *Albany* (NY) *Journal*, October 11, 1833.

26 "During four evenings of this week": "Exhilarating Gas," *Fayetteville* (NC) *Observer*, October 8, 1833, quoting the *Pittsburgh Manufacturer.*

26 "every article for the Sportsman's use": *Matchett's Baltimore Director for 1837* (Baltimore, MD: printed by the author, 1837): 32.

27 For Pearson this was a mistake: Samuel Colt's dealings with Baxter and Pearson can be found in Edwards, *The Story of Colt's Revolver,* 31–38. Additional information on Pearson comes from Robert Pershing, "John Pearson: Gunsmith for Sam Colt," *American Society of Arms Collectors Bulletin* 103 (Spring 2011): 24–32.

27 "I am out of money": Edwards, *The Story of Colt's Revolver*, 36.

27 Exploding subsurface mines: A report on Colt's venture into exploding mines is in Philip K. Lundeberg, *Samuel Colt's Submarine Battery: The Secret and the Enigma* (Washington, DC: Smithsonian Institution Press, 1974). A chapter on Colt and his mines is in Alex Roland, *Underwater Warfare in the Age of Sail* (Bloomington: Indiana University Press, 1978), 134–149.

27 "inhuman system": The quotes of General Sir Howard Douglass and John Quincy Adams attacking the use of mines as unjust come from John S. Barnes, *Submarine Warfare, Offensive and Defensive* (New York: Van Nostrand, 1869), 52, 56.

28 "a magnificent and astonishing spectacle was presented to us": "Colt's Submarine Battery," *Brooklyn Daily Eagle*, August 23, 1842.

28 And on the Potomac River: Reports of the April 1844 ship explosion come from "Correspondence of the Courier," *Charleston* (SC) *Courier*, April 17, 1844; and "Explosive Experiment," *Alexandria* (VA) *Gazette*, April 17, 1844.

3: MASTER OF STEEL

30 According to the tale: The story of the Remington founder and his sons has been a mix of fact and legend since the nineteenth century, often steered by those promoting the company but also stemming from reliable historical accounts. What you read here comes from a mixture of sources, including Alden Hatch, *Remington Arms in American History,* rev. ed. (Remington Arms, 1972); *Papers Read Before the Herkimer County Historical Society During the Years 1896, 1897 and 1898*, compiled by Arthur T. Smith, Secretary of the Society (Herkimer, NY: Citizen, 1899); and *Remington Centennial Historical Souvenir Programme of the Remington Centennial Celebration* (Ilion, NY: August 29, 30, 31, 1916).

31 Eliphalet built a forge: Information on the operation of the forge comes from Roy Marcot, *Remington: "America's Oldest Gunmaker"* (Peoria, IL: Primedia, 1998), 7.

32 "This he welded skilfully": "The Romance of Remington Arms," *A New Chapter in an Old Story* (New York: Remington Arms-Union Metallic Cartridge, 1912).

32 It's a nice story, but: Historians, including ones writing narratives for the company, have rejected the romantic version of the first Remington rifle. See Marcot, *Remington: "America's Oldest Gunmaker,"* 7–8. An improbable early legend had Lite making the entire gun. That "founder story" suited the company's purpose, because it "resonated with American cultural values in the middle part of the nineteenth century, a period that admired young men who took individual initiative, did something well, and achieved success"; Terrence H. Witkowski, "Mythical moments in Remington brand history," *Culture and Organization* 22, no. 1 (2016): 44, 49, citing William Hosley, "Guns, Gun Culture, and the Peddling of Dreams" in *Guns in America: A Reader*, ed. Jan E. Dizard, Robert Merrill Muth, and Stephen P. Andrews, Jr. (New York: New York University Press, 1999), 47–85.

33 fears of a north-south "dismemberment": Peter L. Bernstein, *Wedding of the Waters: The Erie Canal and the Making of a Great Nation* (New York: W. W. Norton, 2005), 63.

34 A Delaware chemical company: Bernstein, *Wedding of the Waters*, 204.

34 On New Year's Day 1828: Remington's purchase and the description of the settlement are in *History of Herkimer County, N.Y.* (New York: F. W. Beers, 1879), 161. For the Remington building's foundation being part of the canal wall, see Erie Canal Museum, Martin Morganstein, and Joan H. Cregg, *Erie Canal: Images of America* (Charleston, SC: Arcadia, 2001), 98.

35 Despite his lifelong devotion to poetry: The description of Remington's personality and attire is from "Ilion and the Remingtons," an address given by

Albert N. Russell before the Herkimer County Historical Society, September 14, 1897; and *Papers Read Before the Herkimer County Historical Society*, p. 81 of the 1897 papers.

35 He dressed well: Remington shared with Abraham Lincoln the trait of storing papers in his top hat. "Hats were important to Lincoln: They protected him against inclement weather, served as storage bins for the important papers he stuck inside their lining, and further accentuated his great height advantage over other men." Harold Holzer, *Lincoln at Cooper Union: The Speech That Made Abraham Lincoln President* (New York: Simon & Schuster, 2004), 101.

35 "long, loose-jointed body": Remington's physical appearance is from Hatch, *Remington Arms in American* History, 7; and Smith, *Papers Read Before the Herkimer County Historical Society*, p. 81 of the 1897 papers.

36 "Remington never smiled again: Hatch, *Remington Arms in American History*, 52–53.

36 Most rifle-barrel-makers: While Remington's firm was a trendsetter in using power-driven hammers in his forge, he was not alone in adopting them. "Not until the 1840s, with the rise of a number of new, more heavily capitalized companies such as Colt, Remington, and Robbins & Lawrence, did highly mechanized operations become commonplace among the largest arms producers." Merritt Roe Smith, *Harpers Ferry Armory and the New Technology: The Challenge of Change* (Ithaca, NY: Cornell University Press, 1980), 326.

36 Whether Remington was really the first to make cast-steel gun barrels has been questioned. Merritt Roe Smith says there is evidence that John Hall experimented with cast steel before Remington did. See *Harpers Ferry Armory and the New Technology*, 217, n. 47. In *Simeon North, First Official Pistol Maker of the United States: A Memoir* (Concord, NH: Rumford, 1913), 187–88, authors S. N. D. North and Ralph H. North claimed that their great-grandfather "was first of the old established armorers to adopt [steel], and was before Remington & Son in the actual delivery of steel barreled guns."

36 He lifted up a floorboard: Dropping bundles on canal boats comes from Codman Hislop, *The Mohawk* (New York: Rinehart, 1948), 303.

36 Some believed a new post office: The discussion of naming the post office can be found in *History of Herkimer County, N.Y.*, 161–62; George A. Hardin, ed., *History of Herkimer County, New York, illustrated with portraits of many of its citizens* (Syracuse, NY: D. Mason, 1893), 201–2; and *Remington Centennial Historical Souvenir Programme of the Remington Centennial Celebration* (Ilion, NY: August 29, 30, 31, 1916).

37 English-born John Griffiths: For the contract with the US Army Ordnance Department, see Howard Michael Madaus and Simeon Stoddard, *The Guns of Remington: Historic Firearms Spanning Two Centuries* (Evington, VA: Biplane

Productions, 1997), 9–10. Griffiths worked at the Harpers Ferry Armory before relocating to Cincinnati in 1841. Smith, Harpers Ferry Armory and the New Technology, 247; George D. Moller, American Military Shoulder Arms, vol. 3 (Albuquerque: University of New Mexico Press, 2011), 117.

4: MOBTOWN

39 "The city pulsated with excitement": Robert V. Remini, *Henry Clay: Statesman for the Union* (New York: W. W. Norton, 1991), 644.

39 He and his twin brother: Information about Winchester family members comes from a variety of sources, including census records, land records, newspapers, and yearly city directories; Laura Trevelyan, *The Winchester: The Gun that Built an American Dynasty* (New Haven, CT: Yale University Press, 2016); and *Winchester Notes*, a genealogical record published privately in 1912 by Fanny Winchester Hotchkiss, p. 3.

40 "always hungry and always cold": George Madis, *The Winchester Era* (Brownsboro, TX: Art and Reference House, 1984), 28.

40 His luck involved another man's death: The Winchester inheritance from a half brother can be found in the Maryland State Archives, Baltimore County Court (Chancery Papers) 1815–1851, MSA C295–1182 (George S. Gibson and James Madison Lankey v. Cyrus Britt [sic] et al., petition to sell lot on Lexington Street), MSA C295–970 (Hannah Winchester v. James Murray, et al., petition to sell lot on Fish Street).

41 "There is not to be found": Hezekiah Niles, "The City of Baltimore," *Weekly Register* 3, no. 3 (1812): 45. The *Register*, published and edited by Niles in Baltimore, was the country's first weekly newsmagazine.

41 some forty-seven thousand other people: Baltimore's population was most likely greater than 47,000 in 1812, since that figure came from the 1810 census.

41 Early nineteenth-century Baltimore: For discussions on early Baltimore, see Sherry H. Olson, *Baltimore: The Building of an American City* (Baltimore, MD: Johns Hopkins University Press, 1980); and Seth Rockman, *Scraping By: Wage Labor, Slavery, and Survival in Early Baltimore* (Baltimore, MD: Johns Hopkins University Press, 2009).

42 Before long, Winchester gave up carpentry: Oliver Winchester's first published advertisement as a clothier appeared in W. G. Lyford's *The Baltimore Address Directory* (Baltimore: Joe Robinson, October 1836), 124.

42 There were other clothing makers in town: About two dozen tailors in downtown Baltimore in 1833 "commonly ke[pt] a general assortment of superfine cloths, cassimeres, and vestings, and they ma[de] suits in the most fashionable style, to order, and at the shortest notice." Charles Varle, *A Complete View of Baltimore, with a Statistical Sketch . . .* (Baltimore: Samuel Young, 1833), 146–48.

43 The first half of the nineteenth century: For American clothing and class, as well as the quote from the British consul, see Daniel J. Boorstin, "A Democracy of Clothing" in *The Americans: The Democratic Experience* (New York: Vintage Books, 1974), 91–100.

43 Baltimore was not always a tranquil town: For a discussion of turbulent Baltimore in the early nineteenth century, see Paul A. Gilje, "The Baltimore Riots of 1812 and the Breakdown of the Anglo-American Mob Tradition," *Journal of Social History* 13, no. 4 (Summer 1980): 547–64.

43 Episodic mobbism had been: Mobbism and the bank riot of 1835 are discussed in Robert E. Shalhope, *The Baltimore Bank Riot: Political Upheaval in Antebellum Maryland* (Urbana: University of Illinois Press, 2009); Francis F. Beirne, *The Amiable Baltimoreans* (New York: E. P. Dutton, 1951), 142–55. Sporadic disorders occurred nationwide at that time, the result of several factors, including "the spread of the factory system and deep changes in the nature of work [that] were already in the process of transforming the demographic landscape and threatening social class relationships." Carl E. Prince, "The Great 'Riot Year': Jacksonian Democracy and Patterns of Violence in 1834," *Journal of the Early Republic* 5, no. 1 (Spring 1985): 5.

43 "the speed and frequency": Beirne, *The Amiable Baltimoreans*, 142.

44 "Seize the moment of excited curiosity": John P. Kennedy, *Memoirs of the Life of William Wirt*, rev. ed., vol. 2 (Philadelphia: Blanchard and Lea, 1856), 360. See also Mary Jo Ignoffo, *Captive of the Labyrinth: Sarah L. Winchester, Heiress to the Rifle Fortune* (Columbia: University of Missouri Press, 2010), 25.

44 "go-to city": Ronald G. Shafer, *The Carnival Campaign: How the Rollicking 1840 Campaign of "Tippecanoe and Tyler Too" Changed Presidential Elections Forever* (Chicago: Chicago Review Press, 2016), 62.

44 Anyone interested in siding with: O. F. Winchester advertised Harrison memorabilia in *American and Commercial Daily Advertiser* (Baltimore, MD), April 30, 1840.

44 "The whole place resembled a fair": Oscar Sebourne Dooley, "The Presidential Campaign and Election of 1844" (Ph.D. diss., Indiana University, 1942), 326–27, citing the *National Intelligencer*, May 2, 1844.

44 "Clay badges hung": Remini, *Henry Clay: Statesman for the Union*, 644.

5: FROM BAYONETS TO GUNS

46 Allen had a company: Discussion of Ethan Allen and his pepperboxes can be found in Lewis Winant, *Pepperbox Firearms*, sp. ed. (Birmingham, AL: Palladium Press, 2001), 27–45; Paul Henry, *Ethan Allen and Allen & Wheelock: their guns and their legacy* (Woonsocket, RI: Andrew Mowbray, 2006); and Harold

R. Mouillesseaux, *Ethan Allen, Gunmaker: His Partners, Patents & Firearms* (Ottawa, ON: Museum Restoration Service, 1973).

47 Smith was born in late 1808: For an account of Cheshire at this time, see Ellen M. Raynor and Emma L. Petitclerc, *History of the Town of Cheshire, Berkshire County, Mass.* (Holyoke, MA: Clark W. Bryan, 1885), 83–111. The Mammoth Cheese is discussed on pp. 86–87.

47 It was served at the White House: The White House was called the President's House when Jefferson lived there.

48 The Armory superintendent who took over: A detailed history of Springfield Armory is in Derwent Stainthorpe Whittlesey, "The Springfield Armory: A Study in Institutional Development" (Ph.D. diss., University of Chicago Department of History, 1920).

48 By 1812 musket production: For increasing division of labor and the tripling of annual musket production at Springfield Armory by 1812, see Michael S. Raber, "Conservative Innovators, Military Small Arms, and Industrial History at Springfield Armory, 1794–1918," *Journal of the Society for Industrial Archeology* 14, no. 1 (1988): 3.

48 Too many guns had performed poorly: Thomas Tyson, "Accounting for Labor in the Early 19th Century: The U.S. Arms Making Experience," *Accounting Historians Journal* 17, no. 1 (June 1990): 48.

48 With all this progress came noise: The quotes about the noise and spark showers come from Jacob Abbott, "The Armory at Springfield," *Harper's New Monthly Magazine* 5, no. 26 (July 1852): 6.

49 "It is immaterial whether": The personal responsibility of workers for substandard products is discussed in Abbott, "The Armory at Springfield," 5–6.

49 "a patient energy": excerpt from an old but undated and unsourced newspaper article in the archives of the Wood Museum, Springfield, MA.

50 He spent twenty-six of January's thirty-one days: Horace Smith's time at Springfield Armory is found in payroll records from July 1827 to December 1840 on microfilm at the Springfield Armory National Historic Site, Springfield, MA.

50 "Vigorous, assertive, stubborn": Whittlesey, "The Springfield Armory: A Study in Institutional Development," 122.

51 Word of this reached Andrew Jackson: Jackson's threat to hang Ripley is from Robert V. Bruce, *Lincoln and the Tools of War* (Indianapolis, IN: Bobbs-Merrill, 1956), 23.

51 Before Horace Smith settled: Much early history of Horace Smith comes from Roy G. Jinks, *History of Smith & Wesson: No Thing of Importance Will Come Without Effort,* 14th printing (North Hollywood, CA: Beinfeld, 2004); and

"Horace Smith & Daniel B. Wesson: The two partnerships of Horace Smith and Daniel Baird Wesson forever impacted the world of firearms" in Roy G. Jinks, *Smith & Wesson Handguns 2002* (Peoria, IL: Primedia Special Interest Publications, 2001), 10–17.

51 "He wore in his belt": Mark Twain, *Roughing It* (Hartford, CT: American, 1872), 23–24.

52 That pepperboxes lacked the accuracy: The supposed female duelists from New York are mentioned in Winant, *Pepperbox Firearms*, 32.

52 In his spare time, Thurber wrote poetry and composeqd hymns: Frances Manwaring Caulkins, *History of Norwich, Connecticut; From Its Possession by the Indians to the Year 1866* (published by the author, 1866), 645.

52 In 1843 he patented an early typewriter: Charles Thurber's explanation of his typewriter's usefulness comes from Specification of Letters Patent No. 3,228, August 26, 1843.

53 Thurber's machine was the first: Jerome Bruce Crabtree, *The Marvels of Modern Mechanism and Their Relation to Social Betterment* (Springfield, MA: King-Richardson, 1901), 612.

6: THE INDENTURED BROTHER

54 His father, Rufus: Rufus Wesson urging Daniel into the shoe business is in "Death of Daniel B. Wesson," *The Iron Age: A Review of the Hardware, Iron, Machinery and Metal Trades* 78, no. 6 (August 9, 1906): 385.

54 Shoemaking had become: Information on the number of localities in Worcester County, including the town of Worcester, where shoes were made in 1837, comes from Blanche Evans Hazard, *The Organization of the Boot and Shoe Industry in Massachusetts Before 1875* (Cambridge, MA: Harvard University Press, 1921), 212; For the importance of the shoe and boot business to Worcester, see Franklin Pierce Rice, *Dictionary of Worcester (Massachusetts) and Its Vicinity with Maps of the City and of Worcester County, First Issue* (Worcester, MA: F. S. Blanchard, 1889), 45.

54 well known in the region: Rufus Wesson information is from William Richard Cutter and William Frederick Adams, *Genealogical and Personal Memoirs Relating to the Families of the State of Massachusetts* IV (New York: Lewis Historical, 1910), 2260; Roy G. Jinks, *History of Smith & Wesson: No Thing of Importance Will Come Without Effort,* 14th printing (North Hollywood, CA: Beinfeld, 2004), 5–6; and Ellery Bicknell Crane, *Historic Homes and Institutions and Genealogical and Personal Memoirs of Worcester County Massachusetts With a History of Worcester Society of Antiquity* vol. 1 (New York: Lewis, 1907), 229–30.

55 His shop in Northborough: The location of Edwin Wesson's shop comes from

Josiah Coleman Kent, *Northborough History* (Newton, MA: Garden City Press, 1921), 162.

55 "The truth is": Edwin Wesson's order book is in the personal collection of Smith & Wesson historian Roy G. Jinks.

55 There were gunsmiths around: One contemporary author referred to the Wesson rifle as "so elegant in appearance" with "unsurpassed accuracy." John Ratcliffe Chapman, *Instructions to Young Marksmen* (New York: D. Appleton, 1848), 11–12.

55 Edwin thought that: In *Instructions to Young Marksmen*, Chapman discusses "gaining" twist in a barrel's rifling and attributes its early use to Edwin Wesson, pp. 137–38. See also *Clark's Illustrated Treatise on the Rifle, Shot-gun and Pistol* (Memphis, TN: H.F. Clark, 1850), 12–13.

55 He did suffer from: "The Smith & Wesson Revolver: Its Invention and Manufacture," *Springfield* (MA) *Republican*, June 18, 1893. The newspaper attributes the comment about Clark's nervousness to Daniel Wesson.

55 His creation was a false muzzle: John D. Hamilton, "Alvan Clark and the False Muzzle," *American Society of Arms Collectors Bulletin*, no. 79 (Fall 1998): 31–37; and Ned H. Roberts, *The Muzzle-Loading Cap Lock Rifle*, rev. ed. (Harrisburg, PA: Stackpole Books, 1952): 66–69.

56 So good was Clark's brainchild: While not as prolific an inventor as Walter Hunt, Alvan Clark was as varied in his talents. He eventually abandoned portrait painting to make telescopes that became world famous for their quality. Some of Clark's portraits are in Washington, DC's National Gallery of Art and Boston's Museum of Fine Arts.s

56 Less than a year after: Alvan Clark, "On Rifle Shooting," is in James J. Mapes, ed., *The American Repertory of Arts, Sciences, and Manufactures 3, no. 3* (New York: W. A. Cox, 1841): 164–69.

56 "I must come soon": Quotes from Alvan Clark's correspondence come from letters in the Edwin Wesson Collection, Harriet Beecher Stowe Center (HBSC), Hartford, CT. His claim to be "the best rifle shot in the world" is from Deborah Jean Warner, *Alvan Clark & Sons, Artists in Optics* (Washington, DC: Smithsonian Institution Press, 1968), 10, quoting Garth Galbraith, "The American Telescope Makers," *Cambridge* (MA) *Chronicle*, March 12, 1887.

57 "Extensive infringements": Clark's published complaint about patent infringers appeared in "Rifle Shooting," *American Turf Register and Sporting Magazine* 13 (New York: The Spirit of the Times, 1842): 244.

57 "Daniel likes to hunt": The letter about Daniel's fondness for working on gunlocks is in the Edwin Wesson Collection (HBSC).

57 Like many master-apprentice: Indentured servitude was not unusual in early

America, though it was waning in the mid-nineteenth century. It was often a form of apprenticeship, as was the case with Benjamin Franklin. A book that discusses indentures and apprenticeships in the evolving American economy is W. J. Rorabaugh, *The Craft Apprentice: From Franklin to the Machine Age in America* (New York: Oxford University Press, 1986).

57 The contract signed: Daniel Wesson's indenture contract is in the HBSC's Edwin Wesson Collection.

58 He was about to turn twenty: The book Daniel Wesson gave to Cynthia Hawes is in the collection of Roy G. Jinks.

60 A pair of riflemen: The challenge was in the Philadelphia *Daily Chronicle*, December 21, 1844.

60 In December 1845: Edwin's brothers' pleas for money are in the Edwin Wesson Collection at HBSC.

7: REVOLUTION IN THE BLOOD

61 "Month after month": Spencer read "The Inventor," a poem he had written, at a banquet in 1917. The transcript of his address is in the Christopher Miner Spencer Collection of the Windsor Historical Society (WHS), Windsor, CT.

61 In fact Hollister had told everyone: Description of Josiah Hollister and his service in the Revolution comes from *The Hollister Family of America; Lieut. John Hollister of Wethersfield, Conn., and His Descendants*, compiled by Lafayette Wallace Case, M.D. (Chicago, IL: Fergus, 1886), 95–96; Percival Hopkins Spencer's application for membership in the Connecticut Society of the Sons of the American Revolution, March 5, 1929; and Mary Philotheta Root, ed., *Patriots' Daughters* (New Haven, CT: Connecticut Chapters, Daughters of the American Revolution, 1904), 220.

62 Spencer, whose childhood nickname: Accounts of Spencer's youth come from an autobiographical sketch at the WHS; "A Conversation with Vesta Spencer Taylor, Interviewed by Dick Bertel in 1965," WHS, Accession No. 1984.63.1, Tape #30a, p. 3; and "Biography of Christopher Miner Spencer," by Vesta Spencer Taylor, WHS, Series II, Box 1.11.

63 By the mid-1840s: Information on manufacturing in Manchester is from Mathias Spiess and Percy W. Bidwell, *History of Manchester Connecticut* (Manchester, CT: Centennial Committee of the Town of Manchester, 1924), 91–103; and William E. Buckley, *A New England Pattern: The History of Manchester, Connecticut* (Chester, CT: Pequot, 1973), 80–94.

64 The Spencers claimed as an ancestor: The Spencer family history, including Crit's chair-making ancestors, is contained in Jack Taif Spencer and Edith Woolley Spencer, *The Spencers of the Great Migration, Volume I, 1300 A.D.—1783 A.D.* (Baltimore: Gateway, 1997), 157–79, 302–3.

8: EYES WEST

66 "War was his element": J. Jacob Oswandel, *Notes of the Mexican War, 1846–47–48* (Philadelphia: 1885), 354.

67 "nothing so much interested me": Samuel H. Walker, "Florida and Seminole Wars: Brief observations on the conduct of the officers, and on the discipline of the army of the United States," July 1, 1840, Texas State Library and Archives Commission, Samuel Hamilton Walker Papers, Box No. 1982/47, Item #43.

67 For whose "cultural and spiritual identity": C.S. Monaco, *The Second Seminole War and the Limits of American Aggression* (Baltimore: Johns Hopkins University Press, 2018), 3.

68 In 1840 Walker risked burning the bridge: Walker's screed against the Army is from "Florida and Seminole Wars . . . ," 10, 11.

68 The Panic of 1837: For a history of the Panic see Alasdair Roberts, *America's First Great Depression: Economic Crisis and Political Disorder after the Panic of 1837* (Ithaca, NY: Cornell University Press, 2012).

69 "Religious insanity": See Ronald L. Numbers and Janet S. Numbers, "Millerism and Madness: A Study of 'Religious Insanity' in Nineteenth-Century America" in Ronald L. Numbers and Jonathan M. Butler, eds., *The Disappointed: Millerism and Millenarianism in the Nineteenth Century* (Knoxville: University of Tennessee Press, 1993), 94–97.

69 "This is the flag": Amanda Beyer-Purvis, "The Philadelphia Bible Riots of 1844: Contest Over the Rights of Citizens," *Pennsylvania History: A Journal of Mid-Atlantic Studies* 83, no. 3 (Summer 2016): 383.

69 A byproduct of these twin forces: The term "Manifest Destiny" is generally attributed to John L. O'Sullivan, influential editor of the *Democratic Review*, in a December 27, 1845, column he wrote for the *New York Morning News*. There he declared that it was "our manifest destiny to overspread and to possess the whole of the continent which Providence has given us for the development of the great experiment of liberty and federated self-government entrusted to us."

70 "The love of chivalric immortal fame still clings to my heart": excerpt from a letter from Walker to Ann Walker, January 22, 1842, Samuel Hamilton Walker Papers, Archives Division, Texas State Library and Archives Commission, Austin, quoted in Jimmy L. Bryan Jr., "The Patriot-Warrior Mystique" in Alexander Mendoza and Charles David Grear, eds., *Texans and War: New Interpretations of the State's Military History* (College Station: Texas A&M University Press, 2012), 118.

71 "I write to inform you I am in good health" excerpt from a letter from Walker to Ann Walker, May 6, 1843. Box. No. 1989/080, Item #10, Samuel Hamilton Walker Papers, Archives and Information Services Division, Texas State Library and Archives Commission. In the letter, Walker said there were

eighteen men executed in what came to be known as the Black Bean Episode. Seventeen were killed at one time, while another man, accused of being the leader, was summarily executed later. See Marilyn McAdams Sibley, ed., *Samuel H. Walker's Account of the Mier Expedition* (Austin: Texas State Historical Association, 1978), 59.

72 "It is much more important that we unite": Henry Clay to John J. Crittenden, December 5, 1843, quoted in Daniel Walker Howe, *What Hath God Wrought: The Transformation of America, 1815–1848* (New York: Oxford University Press, 2007), 706.

72 Two months after the war officially started: Colt's letter to Polk is reproduced in *Sam Colt's Own Record of Transactions with Captain Walker and Eli Whitney, Jr., in 1847,* foreword and notes by John E. Parsons (Prescott, AZ: Wolfe, 1992, originally published by the Connecticut Historical Society, 1949).

73 In his official reports: Taylor's comment that Walker "performed very meritorious service as a spy and partisan" is from J. Reese Fry, *A Life of Gen. Zachary Taylor* (Philadelphia: Grigg, Elliot, 1848), 147.

73 "rapid, untiring, terrible": James C. Wilson, "Address on the Occasion of Removing the Remains of Captains Walker and Gillespie on the Twenty-First of April, A.D. 1856" (San Antonio, TX: Published by a Committee of Citizens), 14–15.

74 "exemplary in his habits": Edmund L. Dana, "Incidents in the Life of Capt. Samuel H. Walker, Texan Ranger," *Proceedings of The Wyoming Historical and Geological Society, For the Year ending February 11, 1882* (Wilkes-Barre, PA: The Society, 1882), 57.

74 "I have so often herd of you": The correspondence between Walker and Colt comes from *Sam Colt's Own Record.*

76 The next day he wrote: Whitney's letter to Colt is in *Sam Colt's Own Record.*

77 "You may expect to see": excerpts from a letter in the Edwin Wesson Collection, Harriet Beecher Stowe Center, Hartford, CT.

9: CRAFTING A WARRIOR'S TOOL

78 In a February 5, 1847, letter: Colt's and Walker's letters to Wesson are in the Edwin Wesson Collection, Harriet Beecher Stowe Center, Hartford, CT.

79 "so that we are all mechanics, a distinction which I am proud to acknowledge": Samuel H. Walker, "Brief observations on the conduct of the officers, and on the discipline of the army of the United States," July 1, 1840, Texas State Library and Archives Commission, Samuel Hamilton Walker Papers, Box No. 1982/47, Item # 43.

80 Maryland politicians: The Maryland General Assembly's resolution can be found in "Laws Made and Passed by the General Assembly of the State of Maryland, At a Session Begun and Held at Annapolis, on Monday, the 28th

Day of December, 1846, and Ended on the 10th Day of March, 1847," Resolution No. 22, passed February 24, 1847, [Maryland] *General Assembly, Session Laws,* Vol. 611, 382.

81 "and all I want now is the arms": Colt's and Walker's correspondence to each other is in *Sam Colt's Own Record of Transactions with Captain Walker and Eli Whitney, Jr. in 1847, Foreword and Notes by John E. Parsons* (Prescott, AZ: Wolfe, 1992, originally published by the Connecticut Historical Society, 1949).

83 Since Thompson was close: Information on Captain Stiles and Thompson's matrimonial offerings comes from Ellen D. Larned, *History of Windham County, Connecticut 1760–1880, Vol. 2* (Worcester, MA: printed by the author, 1880), 535; and Richard M. Bayles, ed., *History of Windham County, Connecticut* (New York: W. W. Preston, 1889), 705–6.

83 For their wedding: *The Barbour Collection of Connecticut Town Vital Records, Thompson 1785–1850* (Baltimore: Genealogical Publishing Co., 2002), 363.

10: WALKER GETS HIS COLTS

85 For that reason Mexicans in the area: That Mexicans feared and hated Walker's dragoons, see *Autobiography of an English Soldier in the United States Army* (New York, NY: Stringer & Townsend, 1854), 222.

85 "Should Capt. Walker come across": J. Jacob Oswandel, *Notes of the Mexican War, 1846–47–48* (Philadelphia: 1885), 198.

85 "an inveterate hatred against the Mexicans": excerpt from a June 20, 1847, entry in the diary of Thomas Barclay, quoted in Allan Peskin, ed., *Volunteers: The Mexican War Journals of Private Richard Coulter and Sergeant Thomas Barclay, Company E, Second Pennsylvania Infantry* (Kent, OH: Kent State University Press, 1991), 110.

85 "Col. Harney says they are the best arm": Samuel Hamilton Walker Papers at the Texas State Library and Archives Commission, Box No. 1982/47, Item #8, SHW to JTW, 10/5/1847. It is printed in Charles T. Haven and Frank A. Belden, *A History of the Colt Revolver and the Other Arms Made by Colt's Patent Fire Arms Manufacturing Company from 1836 to 1940* (New York: Bonanza Books, 1940), 292–93.

86 Colt had even added a touch of artistry: The scenes Colt had engraved around the cylinders of his early revolvers are discussed in Arthur Tobias, *Colt Cylinder Scenes 1847–1851, W. L. Ormsby's Texas Rangers and Comanches Fight, Stagecoach Holdup, and Naval Engagement Scene, Engravings On Samuel Colt's Revolvers* (Los Angeles: published by the author, 2011).

86 "gilding the peaks of the distant mountains": Albert G. Brackett, *General Lane's Brigade in Central Mexico* (Cincinnati, OH: H. W. Derby, 1854), 88.

87 By 11 o'clock: Reports of the battle of Huamantla, including Walker's death,

vary. Descriptions come from several sources, including Oswandel, *Notes of the Mexican War*; Brackett, *General Lane's Brigade in Central Mexico*; Thomas Claiborne Papers, Collection No. 00152, The Southern Historical Collection at the Louis Round Wilson Special Collections Library at the University of North Carolina; *Niles' National Register* of November 27, 1847; Horatio O. Ladd, *History of the War with Mexico* (New York: Dodd, Mead, 1883), 260–64; and G.W.M., "Battle of Huamantla," *Brooklyn* (NY) *Daily Eagle*, December 3, 1850.

87 "They made a most magnificent appearance": Brackett, *General Lane's Brigade in Central Mexico*, 90.

88 There were "shouts, screams, reports of fire arms": Lieutenant William Wilkins, quoted in Joseph Wheelan, *Invading Mexico: America's Continental Dream and the Mexican War, 1846–1848* (New York: Carroll & Graf, 2007), 387.

11: THE FALL OF EDWIN WESSON

90 Although growing in population: The early history of Northborough came from several sources, including Rev. Joseph Allen, *Topographical and Historical Sketches of the Town of Northborough, with the Early History of Marlborough, in the Commonwealth of Massachusetts, Furnished for the Worcester Magazine* (Worcester, MA: W. Lincoln & C. C. Baldwin, 1826); Josiah Coleman Kent, *Northborough History* (Newton, MA: Garden City Press, 1921); Bruce Clouette, *Phase II Historic Properties Survey Town of Northborough, Massachusetts*, prepared for Northborough Historical Commission by Public Archaeology Survey Team, Inc., 2009; and William H. Mulligan, Jr., *Northborough: A Town and Its People, 1638–1975* (Northborough American Revolution Bicentennial Commission, 1977, 1982).

90 Making combs: On comb-making in Northborough, see Victoria Sherrow, *Encyclopedia of Hair: A Cultural History* (Westport, CT: Greenwood, 2006), 92; and Mulligan, *Northborough: A Town and Its People*, 123–24.

91 A group of local lenders: Wesson's borrowing is revealed in the Edwin Wesson Collection, Harriet Beecher Stowe Center (HBSC), Hartford, CT.

91 "most famous manufacturer of rifles": *Scientific American* 3, no. 30 (April 15, 1848): 235.

91 "This is beyond all doubt": "Postcript: By Magnetic Telegraph," *Hartford Daily Courant*, November 9, 1847.

92 Edwin also heard: The gun-fancying British engineer was John Ratcliffe Chapman, author of *Instructions to Young Marksmen* (New York: D. Appleton, 1848). Chapman is credited with being an early maker of telescopic rifle sights.

93 "the inventor is unwilling to dispose": *A Compilation of the Messages and Papers of the Presidents* 6 (New York: Bureau of National Literature, 1897) 2430–31.

93 His claim that the patent didn't infringe on Colt's is documented in the Edwin Wesson Collection, HBSC.

93 "I haven't heard from you": The March 19, 1848, letter from Rufus Jr. is at HBSC.

94 "gaining for our city the credit abroad, of furnishing the most celebrated rifle known in the United States": "Notes by a Man about Town," Number 14, *Connecticut Courant*, February 17, 1849.

95 Information on Wesson's Hartford factory comes from "Death of Edwin Wesson," *Hartford (CT) Weekly Times*, February 3, 1849; auction notice for the factory and its contents published in the *Hartford* (CT) *Daily Courant*, November 14, 1849; and "Notes by a Man about Town," Number 14, *Hartford* (CT) *Daily Courant*, January 30, 1849.

95 Wesson's financial documents and correspondence are in the Edwin Wesson Collection at HBSC; in the personal collection of Roy G. Jinks, retired Smith & Wesson historian; and at the Connecticut Historical Society, Hartford, CT.

95 "Now what scene can be more captivating": "Notes by a Man about Town," Number 10, *Hartford* (CT) *Daily Courant*, January 5, 1849.

95 In announcing Edwin's death: "Death of Edwin Wesson," *Hartford* (CT) *Weekly Times*, February 3, 1849, and *Scientific American* 4, no. 21 (February 10, 1849): 162.

95 "had Divine Providence permitted him to live": "Death of Edwin Wesson," *Hartford* (CT) *Weekly Times*, February 3, 1849.

96 These guns are almost universally: The bondmen's report is in the Estate of Edwin Wesson file at HBSC.

97 "Thus ended": Edwin Wesson's order book, which is in the Roy G. Jinks collection.

12: THE WRATH OF SAMUEL COLT

98 the Wesson & Leavitt revolver was the ideal product: Information on Wesson & Leavitt revolvers can be found in Frank M. Sellers and Samuel E. Smith, *American Percussion Revolvers* (Ottawa, ON: Museum Restoration Service, 1971), 90–92.

98 On March 5, 1850: For a discussion of the Massachusetts Arms Company, see L. W. Jones, "Handguns of the Massachusetts Arms Co.," *American Society of Arms Collectors Bulletin* 37 (Fall 1977): 13–19.

101 "strikingly handsome man": "Edward N. Dickerson Dead," *New-York Tribune,* December 13, 1889.

101 "great galvanic battery of human oratory": "Trial of Albert J. Tirrell for the murder of Maria A. Bickford!", *Boston Daily Mail*, March 28, 1846.

101 Four years earlier: For a discussion of Rufus Choate's sleepwalking defense,

see Karen Abbott, "The Case of the Sleepwalking Killer," smithsonianmag.com, April 30, 2012. www.smithsonianmag.com/history/the-case-of-the-sleepwalking-killer-77584095/

102 Two teams of lawyers squared off: Testimony, arguments of counsel, and opinion of the judge come from Robert M. Patterson, *Samuel Colt v. The Mass. Arms Company* (Boston: White & Potter, 1851).

102 He didn't mention Colt's family connections: The Ellsworth letter to Christopher Colt is in the Samuel Colt Papers, MS 28415, Box 1, Folder 1, Connecticut Historical Society, Hartford, CT.

102 And Dickerson did not mention: Rufus Porter's release of his revolver rights to Colt is in RG 103, Records of The Colt Patent Fire Arms Manufacturing Company Legal File, Series III-IV, Complaints and Law Suits—Patents, Box 45, Connecticut State Library, Hartford, CT.

106 "broadly cultured lawyer": William D. Bader and Roy M. Mersky, "Justice Levi Woodbury: A Reputational Study," *Journal of Supreme Court History* 23, no. 2 (1998): 129–42.

106 In Woodbury's view: America's mission: is from Levi Woodbury, "The Annual Address Delivered before The National Institute in the Hall of the House of Representatives, January 15, 1845" (Washington, DC: J. and G. S. Gideon, 1845): 25.

107 There would be no more: Promptly after Colt's patent expired in 1857, the Massachusetts Arms Company resumed making a revolver with a cylinder that locked up like a Colt. "It is a sweet little instrument for knocking holes into people," the *Springfield* (MA) *Republican* advised its readers on March 26, 1857.

13: CONQUERING THE EVIL COLLAR

110 The other watershed event: Winchester's "Improvements in the Method of Cutting and Fitting Shirts," Patent No. 5,421, was issued on February 1, 1848.

111 The year before the Winchesters resettled there: On clocks and mass production, see David A. Hounshell. *From the American System to Mass Production, 1800–1932: The Development of Manufacturing Technology in the United States* (Baltimore: Johns Hopkins University Press, 1984), 51–61.

111 The Industrial Revolution may have brought: On New Haven carriage-building, see Ben Ford, "The Cruttenden Carriage Works: The Development and Decline of Carriage Production in New Haven, Connecticut," *Journal of the Society for Industrial Archeology* 38, no. 1 (2012): 55–74; Richard Hegel, *Carriages from New Haven: New Haven's Nineteenth-Century Carriage Industry* (Hamden, CT: Archon Books, 1974); Preston Maynard and Marjorie B. Noyes, eds., *Carriages and Clocks, Corsets and Locks, The Rise and Fall of an Industrial City—New Haven, Connecticut* (Lebanon, NH: University Press of New England, 2004); and

Rollin G. Osterweis, *Three Centuries of New Haven, 1638–1938* (New Haven, CT: Yale University Press, 1953), 251–52.

112 The Baltimore shirtmaker had come to New England: On Winchester and shirts in New Haven, see Edward E. Atwater, *History of the City of New Haven to the Present Time* (New York: W. W. Munsell, 1887), 628; and Robert B. Gordon, "Industrial Archaeology of New Haven's Harborside Area," in Maynard and Noyes, *Carriages and Clocks,* 29.

112 "Indeed," observed a journal of commerce: Freeman Hunt, *Merchants' Magazine, and Commercial Review* 3, no. 4 (October 1840): 308. Also see Michael Zakim, *Ready-Made Democracy: A History of Men's Dress in the American Republic, 1760–1860* (Chicago: University of Chicago Press, 2003), 55.

113 For Davies clothing had been: Information on Davies comes in large part from Edwin T. Freedley, ed., *Leading Pursuits and Leading Men: A Treatise on the Principal Trades and Manufactures of the United States* (Philadelphia: Edward Young, 1856), 141–44, 147–48.

113 When he was two years old: On Luke Davies, see Frederick Converse Beach, ed., *The Encyclopedia Americana,* vol. 5 (New York: Americana, 1904).

113 His reputation growing: The prize Luke won is mentioned in *Niles' Weekly Register* of October 24, 1829.

113 By 1837 when the father retired: The productivity of the Davies business comes from "American Institute, Stocks," *New York Daily Herald,* October 24, 1837.

113 Winchester was among the early mass-producers of shirts. Claudia B. Kidwell and Margaret C. Christman, *Suiting Everyone: The Democratization of Clothing in America* (Washington, DC: Smithsonian Institution Press, 1974), 55.

113 Shirts "Of Linen": advertisements in Massachusetts and Vermont newspapers in the spring of 1849.

114 "Patent Shoulder Seam Shirts, made to order": *Matchett's Baltimore Director for 1853–54.*

114 "NOTICE TO THE LADIES": the *Baltimore Sun,* October 30, 1851.

114 In 1851 guns still weren't on Oliver Winchester's mind: Winchester's horticultural exploits are found in the *Annual Report of the New Haven County Horticultural Society for the Year 1851* (New Haven, CT: B. L. Hamlen, printer, 1851).

14: TO LONDON 'MIDST IRON AND GLASS

116 "an enchanted pile": *Speech of Mr. Disraeli on the Financial Policy of the Government, Made in the House of Commons. June 30, 1851* (London: Thomas Lewis, 1851), 15.

116 A letter to the religious British newspaper: Geoffrey Cantor, *Religion and the Great Exhibition of 1851* (Oxford: Oxford University Press, 2011), 181.

117 In front of the opening-day crowd: Queen Victoria's comments and "a flourish

of trumpets" are from Joseph Irving, *The Annals of Our Time: A Diurnal of Events, Social and Political, Which Have Happened In, or Had Relation to, the Kingdom of Great Britain, from the Accession of Queen Victoria to the Opening of the Present Parliament* (London: Macmillan, 1869), 203–4.

117 Wrote Charlotte Brontë: excerpt from a letter she wrote to her father on June 7, 1851, which can be found in Clement Shorter, *The Brontës Life and Letters*, vol. 2 (London: Hodder and Stoughton, 1908), 215–16.

118 "certainly not very interesting": The Queen's Exhibition Journal in C. R. Fay, *Palace of Industry, 1851: A Study of the Great Exhibition and Its Fruits* (Cambridge, UK: Cambridge University Press, 1951, first paperback edition 2010), 54.

118 "By packing up the American articles": Quotes from *Punch* can be found in Marcus Cunliffe, "America at the Great Exhibition of 1851," *American Quarterly* 3, no. 2 (Summer 1951): 119–20.

119 "The most popular and famous invention": *The Times* (UK), June 9, 1851.

119 "on the Texan frontier, and on the several routes to California": quoted in John R. Davis, *The Great Exhibition* (Thrupp, UK: Sutton, 1999), 161.

119 "Amongst the most humiliating circumstances in the recent war at the Cape": the *Maidstone* (UK) *Gazette,* quoted in Charles T. Rodgers, *American Superiority at the World's Fair* (Philadelphia: John J. Hawkins, 1852), 31.

120 She "stands exposed to the gaze": Sculptor Hiram Powers's description of his *Greek Slave* is quoted in Henry T. Tuckerman, *Book of the Artists* (New York: G. P. Putnam & Sons, 1867), 285.

120 "The expenditure of months": *Great Exhibition of the Works of Industry of All Nations, 1851, Official Descriptive and Illustrated Catalogue Volume III* (London: W. Clowes and Sons, 1851), 1431.

121 "The Great Exhibition has done more than any thing": *The North American Review*, vol. 74, no. 154 (January, 1852), 199.

121 "Diffuse, then, knowledge": *De Bow's Review of the Southern and Western States* 9—New Series, vol. 1 (New Orleans, LA: J. D. B. De Bow, 1850): 271.

121 "Bowie knives in profusion": The Queen's Exhibition Journal, quoted in Fay, *Palace of Industry,* 52.

122 He also came bearing gifts: Prince Albert's revolvers in their presentation box are in the Royal Collection Trust. www.rct.uk/collection/71649/pair-of-revolvers.

123 "Standing on American ground": William Edward Baxter, *America and the Americans* (London: Geo Routledge & Co., 1855), 17, quoted in Joshua Taylor, "'I Think of the Future': The Long 1850s and the Origins of the Americanization of the World." (Ph.D. diss., University of South Florida, 2019), 16. Taylor's thesis is that American ascendance to industrial prominence began in

the 1850s and was not a sudden leap decades later, as some have argued. The 1850s was also a period of rapid innovation in American arms.

123 "[I]f I cant be first": excerpt from a letter Colt wrote to his half brother, William, on July 21, 1844, quoted in William B. Edwards, *The Story of Colt's Revolver: The Biography of Col. Samuel Colt* (New York: Castle Books, 1957), 153.

15: PERFECTION IN THE ART OF DESTRUCTION

124 his only combat having been in court: For Colt assaulting a lawyer, see "In the Blood," *Morning News* (New London, CT), August 4, 1845.

124 He hired a New York military clothier: Colt's uniform is at the Connecticut State Library's Museum of Connecticut History, Hartford, CT.

125 "Among all the objects of interest": "Testimonials of the Usefulness of Colt's Repeating Fire-arms from European Journals and Officers of the British Army and Navy," extracts from a pamphlet issued in 1852 by the Colt Company, reproduced in Charles T. Haven and Frank A. Belden, *A History of the Colt Revolver and the Other Arms Made by Colt's Patent Fire Arms Manufacturing Company from 1836 to 1940* (New York: Bonanza Books, 1940), 327–35.

125 "Experience has shown": Quotes from Colt's talk to the Institution of Civil Engineers come from Samuel Colt, *On the Application of Machinery to the Manufacture of Rotating Chambered-Breech Fire-Arms, and Their Peculiarities*, 3rd ed. by Charles Manby (London: William Clowes and Sons, 1855). Comments attributed to Abbott Lawrence are quotes from the document that paraphrased what Lawrence said.

126 In case anyone was foolish enough: Dickerson's notice to the gun trade is quoted from Herbert G. Houze, *Samuel Colt Arms, Art, and Invention* (New Haven, CT: Yale University Press, 2006), 125. An original can be found in Samuel Colt Papers, Box 7, Connecticut Historical Society, Hartford, CT.

128 Colt was eager to have his revolvers: Colt's letter to Perry is quoted in Steven Lubar, "In the Footsteps of Perry: The Smithsonian Goes to Japan," *The Public Historian* 17, no. 3 (July 1995): 37.

129 But nothing could compare: The Dragoon that Colt gave to the sultan is now at the Metropolitan Museum of Art in New York City. The tsar's Dragoon is at the State Hermitage Museum, Saint Petersburg, Russia.

129 Thanks to Seymour: The tsar coming to the Winter Palace to meet Colt and information about the failed attempt to smuggle revolvers is from Joseph Bradley, *Guns for the Tsar: American Technology and the Small Arms Industry in Nineteenth-Century Russia* (Dekalb: Northern Illinois University Press, 1990), 56–57.

129 In a leased three-story brick building: *Household Words. A Weekly Journal. Conducted by Charles Dickens* 9, no. 218 (May 27, 1854): 354–56.

130 "SHIRT MAKERS WANTED": the *Hartford (CT) Daily Courant*, September 22, 1848, and the *Semi-Weekly Eagle* (Brattleboro, VT), November 10, 1851.

131 At first Winchester was skeptical: The quotes from Nathaniel Wheeler come from George Madis, *The Winchester Era* (Brownsboro, TX: Art and Reference House, 1984), 29, 31. Other sources say that it was the wife of the inventor Allen B. Wilson who demonstrated the machine to Winchester or that it was Wilson, not Wheeler, who made the pitch.

131 the first manufacturer to use: That Winchester & Davies were the first to use sewing machines to make shirts comes from "Romance of the Shirt Trade," *The Haberdasher* 69, no. 4 (April 1919): 68.

131 From a workforce of two hundred: Details of the production at Winchester & Davies come from Edwin T. Freedley, ed., *Leading Pursuits and Leading Men: A Treatise on the Principal Trades and Manufactures of the United States* (Philadelphia: Edward Young, 1856), 141–44, 147–48.

131 "When we look at the progress made in Sewing Machines": *Scientific American* 7, no. 44 (July 17, 1852): 349.

132 "[W]e firmly believed that some things would never be done by any fingers except human": Henry Ward Beecher, "Our First Experience with a Sewing Machine," *The Massachusetts Teacher (1858–1871)* 13, no. 1 (January 1860): 12–15.

132 In late June 1852: Information on the New Haven Shirt Manufactory party and Oliver Winchester's address to workers are in *Festival at the New Haven Shirt Manufactory, June 25*th*, 1852* (New Haven, CT: J. H. Benham, 1852).

16: EVOLUTION

134 breakthrough inventions that receive patents: "The history of significant innovation in this country is, contrary to popular myth, a history of incremental improvements generally made by a number of different inventors at roughly the same time." Mark A. Lemley, "The Myth of the Sole Inventor," *Michigan Law Review* 110, no. 5 (2012): 760. On pages 718–720, Lemley discusses Whitney and Morse.

135 A full decade before the Patent Office: The "magazine gun" that preceded Hunt's Volition Repeater is Patent No. 1,084, February 20, 1839.

136 Jennings simplified the mechanism: Descriptions of what Jennings did with the Volition Repeater can be found in Joseph G. Bilby, *A Revolution in Arms: A History of the First Repeating Rifles* (Yardley, PA: Westholme, 2006), 54–56; R. Bruce McDowell says that Walter Hunt and Jennings worked together almost from the beginning. See his *Evolution of the Winchester* (Tacoma, WA: Armory, 1985), 1–4.

136 New Yorker Courtlandt Palmer was that someone: Sources for information on

Palmer include his obituaries in the *New York Times*, *New York Tribune*, and the *New York Daily Herald*, all of May 12, 1874. Cornelius Vanderbilt's view of Palmer is found in T. J. Stiles, *The First Tycoon: The Epic Life of Cornelius Vanderbilt* (New York: Knopf, 2009).

137 The stories of the Robbins & Lawrence Armory: An account of Robbins & Lawrence and its significance in industrial development can be found in Joseph Wickham Roe, *English and American Tool Builders* (New York: McGraw-Hill, 1926), 186–201; and Ross Thomson, "The Firm and Technological Change: From Managerial Capitalism to Nineteenth-Century Innovation and Back Again," *Business and Economic History* 22, no. 2 (Winter 1993): 118–19.

137 Richard Lawrence was nine years old: Lawrence's account of his life, including his work on Story's rifle, is Appendix A in Roe, *English and American Tool Builders*.

139 Then they built their factory: Building the Robbins & Lawrence factory is discussed in John P. Johnson, "Robbins & Lawrence Armory (American Precision Museum)," Historic American Engineering Record No. VT-39, National Park Service, Washington, DC, 2009.

141 Whaling was also a pioneer in high-risk, high-reward: Whaling as early venture capitalism is discussed in "Fin-tech: Before there were tech startups, there was whaling," *The Economist*, December 30, 2015.

141 Far better from the whalers' point of view: For a discussion of whaling guns, see Robert E. Hellman, "Whaling Tools in the Nantucket Whaling Museum," *Historic Nantucket* 47, no. 3 (Summer 1998): 16–17; and Lance E. Davis, Robert E. Gallman, and Karin Gleiter, *In Pursuit of Leviathan: Technology, Institutions, Productivity, and Profits in American Whaling, 1816–1906* (Chicago: University of Chicago Press, 1997), 288–94.

141 A fellow Norwich gunsmith: James Temple Brown discussed Oliver Allen and his bomb lance in *The Whale Fishery and Its Appliances* (Washington, DC: Government Printing Office, 1883), 15–16, 58.

142 Among them was Oliver Allen: Allen's relocation to California is mentioned in an obituary for his son in the *Norwich* (CT) *Bulletin*, March 19, 1921.

17: A NEW ORGAN OF DESTRUCTIVENESS

144 Exactly when and where Smith first met Wesson: Information on Smith & Wesson's early years, as well as information on the founders, comes from a variety of sources, including Roy G. Jinks, *History of Smith & Wesson: No Thing of Importance Will Come Without Effort*, 14th printing (North Hollywood, CA: Beinfeld, 2004).

147 "We have seen and fired a pistol": attributed to the *New Haven* (CT) *Palladium* in the *Washington* (AR) *Telegraph*, March 21, 1855, and other newspapers.

148 When a shipment of Volcanics arrived by steamship: The Charleston hardware merchant who offered Volcanics for sale was J. Van Winkle. His advertisement appeared in the *Charleston Daily Courier*, February 14, 1855.

148 "This is a new 'organ of destructiveness'": "City Matters—Volcanic Repeating Pistol," *Milwaukee Daily Sentinel*, March 30, 1855.

148 A couple of months later, a prominent hardware dealer: The Nazro advertisements appeared in the *Milwaukee Daily Sentinel.*

149 "a wholly unnecessary violation": Report on the New Haven anti-Sunday-mail meeting in which Winchester joined a lobbying committee is in "Anti-Sunday Mail Meeting in New Haven," *Hartford* (CT) *Daily Courant*, March 30, 1853.

150 As part of the deal to take over everything: Information on the Volcanic Repeating Arms Company and the New Haven Arms Company comes from a variety of sources, including D. H. Veader and A. W. Earle, *The Story of the Winchester Repeating Arms Company*, 1918, Winchester Repeating Arms Company Archives Collection, MS 20, McCracken Research Library, Buffalo Bill Center of the West, Cody, WY; Herbert G. Houze, *Winchester Repeating Arms Company: Its History & Development from 1865 to 1981* (Iola, WI: Krause, 1994 and 2004); Edmund E. Lewis and Stephen W. Rutter, *Volcanic Firearms: Predecessor to the Winchester Rifle* (Woonsocket, RI: Andrew Mowbray, 2011); and "The Story of the Rise of a Great Industry," *Winchester Record* 1, no. 14 (February 14, 1919).

150 In late 1855 it sent an agent: Joseph Storrs's correspondence with Samuel Talcott is in MS 20.67.29 at the McCracken Research Library, Buffalo Bill Center of the West, Cody, WY.

18: IN KING COLT'S SHADOW

154 Rollin White was only about ten: The account of White's first exposure to a breech-loading gun and working with his brother in Vermont comes from his June 9, 1862, deposition in *Rollin White et al v. Ethan Allen et al* (hereinafter *White v. Allen*), Circuit Court of the United States, Massachusetts District (Boston: Alfred Mudge & Son, 1863), 188.

154 White possessed both property and influence: John Adams Vinton, *The Richardson Memorial, Comprising a Full History and Genealogy of the Posterity of the Three Brothers, Ezekiel, Samuel, and Thomas Richardson* (Portland, ME: Brown Thurston, 1876), 369.

154 He owned a pew: Josiah White's property at the time of his death in 1852 is from ancestry.com, *Vermont, U.S., Wills and Probate Records, 1749–1999*, Provo, UT, original data from Vermont County, District and Probate Courts.

www.ancestry.com/imageviewer/collections/9084/images/004331124_00153?backlabel=ReturnSearchResults&queryId=f030d5b5250a426c106c8e6b3e3845b1&pId=92384

155 Josiah figured that by cutting: White and his opening a trench to Lime Pond is mentioned in *A History of Williamstown Vermont 1781–1991*, Vermont Bi-Centennial Year project of the Williamstown Historical Society, 8, 81. Descriptions of the town are on pp. 8, 12, 80–81.

156 "I suggested to my brother": Rollin White's account of his employment and gunmaking experiments comes from *White v. Allen*, 187–89, 203. Testimony from White and Ferdinand Steele on experiments with the bored-through cylinder comes from *White v. Allen*, 190–95, 213–14.

159 He considered showing his invention to Samuel Colt: Attorney and gun collector/researcher Robert Swartz makes the case that White lied about why he delayed his patent application in order to convince the judge that he had not dawdled. See "A Deplorable Madness—Part I," *Arms Heritage Magazine* 5, no. 2 (April 2015); and "A Deplorable Madness—Part II," *Arms Heritage Magazine* 5, no. 3 (June 2015).

159 On April 3, 1855: The same day that Rollin White received Patent No. 12,648, he also received Patent No. 12,649 for a front-loading revolver.

159 By all accounts Samuel Colt's London factory: A detailed examination of Colt's English venture is Joseph G. Rosa, *Colonel Colt, London: The history of Colt's London firearms, 1851–1857* (London: Arms & Armour, 1976).

160 There were several possible culprits: Some in Britain complained about Colt's failure to make truly interchangeable parts is in Robert A. Howard, "Interchangeable Parts Reexamined: The Private Sector of the American Arms Industry on the Eve of the Civil War, *Technology and Culture* 19, no. 4 (October 1978): 643.

160 new home for Samuel Colt's revolver business: A detailed description of the Colt factory in the mid-1850s—with a fawning treatment of the man—is in "Repeating Fire-Arms: A Day at the Armory of 'Colt's Patent Fire-Arms Manufacturing Company,'" *United States Magazine* 4, no. 3 (March 1857): 221–49.

160 "Emperor of the South Meadows": the *Hartford* (CT) *Daily Courant*, August 11, 1855.

161 In a letter late in 1855: McIntire's and Keller's letters to Colt are quoted in Herbert G. Houze, *Samuel Colt: Arms, Art, and Invention* (New Haven, CT: Yale University Press, 2006), 116. Keller later sold his services to Rollin White in both *White v. Allen* and White's attempt to have his patents extended. Congressional Record, Proceedings and Debates of the Forty-Third Congress, First Session, May 22, 1874, pp. 4179–80.

162 "I notice in a patent granted to you": D. B. Wesson's letter is quoted in Roy G. Jinks, *History of Smith & Wesson: No Thing of Importance Will Come Without Effort,* 14th printing (North Hollywood, CA: Beinfeld, 1977), October 2004, 34–35.

19: THE SMITH & WESSON MOMENT

163 No one had leapt at the chance: Rollin White patent claims appeared in *Scientific American* 10, no. 31 (April 14, 1855): 246.

163 "PRINTED FABRICS": advertisement in the October 29, 1856, issue of the *Hartford* (CT) *Daily Courant.*

163 The document they all signed: The licensing agreement can be found in "Message of the President of the United States, returning Bill S. No. 273, entitled 'An act for the relief of Rollin White,' with his objections," 41st Congress, 2nd Session, Ex. Doc. No. 23, 5–6.

164 the partners opened a fresh set of books: The beginning of the new Smith & Wesson company is discussed in Roy G. Jinks, *History of Smith & Wesson: No Thing of Importance Will Come Without Effort,* 14th printing (North Hollywood, CA: Beinfeld, 1977), October 2004, 37.

165 Below then William L. Wilcox: The description of Wilcox's business comes from an advertisement in M. Bessey, *Bessey's Springfield Directory for 1854–1855* (Springfield, MA: published by the author, 1854), 156.

165 "They will put in an engine": "New Operations in Business and Property," *Springfield* (MA) *Republican*, March 26, 1857.

165 Smith & Wesson advertised their new gun: an advertisement reprinted in Jinks, *History of Smith & Wesson*, 39.

166 "I was armed to the teeth": Mark Twain, *Roughing It* (Hartford, CT: American, 1872), 22–23.

166 where he offered dresses for all occasions: "for breakfast, and dinner": excerpt from several advertisements that Rollin White & Co. placed in the *Hartford* (CT) *Daily Courant* in August and September 1857.

167 During his sojourn abroad: Colt's dealings with Marcotte are from Phillip M. Johnston, "Dialogues between Designer and Client: Furnishings Proposed by Leon Marcotte to Samuel Colt in the 1850s," *Winterthur Portfolio* 19, no. 4 (Winter 1984): 264–66. "robust, sturdy, and conservative" is on 275.

20: OF SILK AND STEEL

169 In the early nineteenth century: On Connecticut's silkworm boom, see Bob Wyss, "Connecticut's Mulberry Craze," August 30, 2020, https://connecticuthistory.org/connecticuts-mulberry-craze/; Patrick Skahill, "The Cheney Brothers' Rise in the Silk Industry," February 9, 2013, https://connecticuthistory.org/the-cheney-brothers-rise-in-the-silk-industry/; and

H.H. Manchester, *The Story of Silk and Cheney Silks* (South Manchester, CT: Cheney Brothers, 1916, rev. 1924).

170 Owen had a good friend in Frank Cheney: On the Cheneys in general, as well as their relationship with the Spencers, see Alice Farley Williams, *Silk and Guns: The Life of a Connecticut Yankee, Frank Cheney, 1817–1904* (Manchester, CT: The Manchester Historical Society, 1996).

170 "I remember him in the little old mill": excerpt from a letter Spencer wrote to one of Frank Cheney's daughters on March 21, 1904, Christopher Miner Spencer Collection, Box 1, Folder 2, Windsor Historical Society (WHS), Windsor, CT.

170 "the ruling spirit": from "A Conversation with Vesta Spencer Taylor, Interviewed by Dick Bertel in 1965," WHS, Accession No. 1984.63.1, Tape #30a, 4.

170 An introduction was made: Spencer's work history comes from "A Conversation with Vesta Spencer Taylor," 3–4; and Roy M. Marcot, *Spencer Repeating Firearms* (Rochester, NY: Rowe, 1983, 1990), 12–13.

171 "I go to sleep thinking about them": Dr. W. A. Bartlett, "An Appreciation of [Christopher] Spencer prepared for the Windsor Historical Society," undated, Spencer Collection, Series II, Box 1.11, WHS.

172 The Cheneys detested slavery: Evidence of the Cheneys' abolitionist leanings comes from "The Underground Railroad," a paper presented by Frank Woodbridge Cheney at the Hartford Monday Evening Club, February 11, 1901, preserved in an unpublished family history, Connecticut Historical Society, Hartford, CT.

173 It was while he was working for the Cheneys: Spencer's handwritten account of his developing a repeating gun is in the Christopher Miner Spencer Collection, Series IA, Correspondence Box 1.2, WHS.

173 He didn't say why he opted: A professor of mechanical and manufacturing engineering and the head curator of the Royal Armouries in Leeds, England, concluded that Spencer probably put the magazine in the butt stock rather than under the barrel to avoid infringing patents descending from Walter Hunt's Volition Repeater, which Winchester controlled. David Williams and Peter Smithurst, "Christopher Spencer: the manufacturing technology of his repeating rifle," *Arms & Armour* 1, no. 2 (2004): 165–82.

21: A GATHERING RUSH FOR WEAPONRY

176 Now on his one-day stop in Hartford: Abraham Lincoln's visit to Hartford is recounted in Daniel D. Bidwell, "Lincoln in Hartford," in William Hayes Ward, ed., *Abraham Lincoln: Tributes from his Associates—Reminiscences of Soldiers, Statesmen and Citizens* (New York: Thomas Y. Crowell, 1895), 182–84; Doris Kearns Goodwin, *Team of Rivals: The Political Genius of Abraham Lincoln* (New York: Simon & Schuster, 2006), 232–34; John Niven, *Gideon Welles: Lincoln's Secretary*

of the Navy (Baton Rouge: Louisiana State University Press, 1973), 287–89; and J. Doyle DeWitt, *Lincoln in Hartford*, a privately printed pamphlet in support of the activities of the Connecticut Civil War Centennial Commission, published in 1960, no place of publication listed.

176 The two men sat together: There are differing accounts of when Lincoln and Gideon Welles got together at the bookstore. Some say it was the day after the lecture and that Lincoln didn't arrive in Hartford until shortly before he gave his speech on the evening of March 5.

177–8 That night, a "pressed down, shaken together": this and "in every way large" are from "Republican Mass Meeting at the City Hall!" and "Mr. Lincoln," *Hartford* (CT) *Evening Press*, March 6, 1860. The inspiration for the language in the former quote was the New Testament, Luke 6:38: "Give, and it shall be given unto you; good measure, pressed down, and shaken together, and running over, shall men give into your bosom."

177 After Lincoln finished: The Wide Awakes are the subject of Jon Grinspan's "'Young Men for War': The Wide Awakes and Lincoln's 1860 Presidential Campaign," *Journal of American History* 96, no. 2 (September 2009): 357–378.

178 The guns coming from: The statistics on Colt's and Sharps's production come from Felicia Johnson Deyrup, *Arms Makers of the Connecticut Valley: A Regional Study of the Economic Development of the Small Arms Industry, 1798–1870* (Menasha, WI: George Banta, 1948), 220, Appendix A, Table 4.

178 The old system of small makers: The growth of private, independent arms makers in the decades before the Civil War is discussed in Deyrup, *Arms Makers of the Connecticut Valley*, 120–32.

179 People moving west were encouraged: The handbook advising those journeying into the continent to be armed with Colt revolvers is Randolph B. Marcy, *Prairie Traveler, a Hand-book for Overland Expeditions* (New York: Harper & Brothers, 1859), 41–43.

180 In the first three years: Information on production and engineering changes to Smith & Wesson Model 1 revolvers is from David R. Burghoff, "Smith & Wesson Rarities, 1854–1900," *American Society of Arms Collectors Bulletin*, no. 56 (Spring 1987): 9–11.

180 In late 1859 they built: Description of Smith & Wesson's new factory on Stockbridge Street is from the *Springfield* (MA) *Republican*, January 20, 1860. Information on the company ball comes from the *Springfield* (MA) *Republican*, December 26, 1859, and December 30, 1859.

181 By the beginning of March: The orders Smith & Wesson received from South America and elsewhere were reported in the *Springfield Republican*, March 1, 1860.

181 In 1858 Remington added: The cane gun is discussed in Elliot L. Burka, "Remington Rifle Cane," *American Society of Arms Collectors Bulletin*, no. 85 (Spring 2002): 1–10.

182 "I consider the Volcanic Repeating Pistol: The Volcanic broadsheet is reproduced in Edmund E. Lewis and Stephen W. Rutter, *Volcanic Firearms: Predecessor to the Winchester Rifle* (Woonsocket, RI: Andrew Mowbray, 2011), 124.

182 "the greatest invention of the age": excerpt from an advertisement in the *Rock Island (IL) Argus*, January 23, 1856.

183 It took him a while: On the cartridge invented by B. Tyler Henry, see R. Bruce McDowell, *Development of the Henry Cartridge and Self-Contained Cartridges for the Toggle-Link Winchesters* (Metuchen, NJ: A.M.B., 1984).

183 "These arms," claimed the company's Baltimore agent: from an advertisement in the Baltimore *Daily Exchange*, February 18, 1861.

183 "the most powerful and most effective weapon ever invented": from an advertisement that appeared several times in the *Baltimore Sun* in February 1861.

22: CALL TO ARMS, ANY ARMS

184 "Run the armory night and day": Colt's letter is quoted in James L. Mitchell, *Colt: A collection of letters and photographs about the man, the arms, the company* (Harrisburg, PA: Stackpole, 1959), 75–76.

185 As it did when the Revolution began: For early arms shortages and buying guns from Europe, see Carl L. Davis, *Arming the Union: Small Arms in the Civil War* (Port Washington, NY: National University Publications, 1973), 45–80.

186 "Permit me to introduce": A photocopy of Buckingham's letter to Cameron is in the author's collection.

186 In a letter to John A. Dahlgren: A copy of Welles's letter to Dahlgren is from the collection of firearms author Andrew F. Lustyik.

186 That friend was Abraham Lincoln: The close relationship between Dahlgren and Lincoln is discussed in Robert J. Schneller, Jr., *A Quest for Glory: A Biography of Rear Admiral John A. Dahlgren* (Annapolis, MD: Naval Institute Press, 1996), 183–89.

186 Four days after the Navy secretary: Dahlgren's letter to Harwood is reproduced in Roy M. Marcot, *Spencer Repeating Firearms* (Rochester, NY: Rowe, 1983, 1990), 28.

187 "Your country needs you": Ripley is quoted in George W. Cullum, *Biographical Register of the Officers and Graduates of the U. S. Military Academy at West Point, N.Y.*, 3rd ed., vol. 1 (Boston: Houghton, Mifflin, 1891), 122.

187 No need for "newfangled gimcracks": Robert V. Bruce, *Lincoln and the Tools of War* (Indianapolis, IN: Bobbs-Merrill, 1956), 112.

187 Ripley's resistance to new technology earned him the nickname: Allan Nevins,

The War for the Union, vol. 1, The Improvised War, 1861–1862 (New York: Charles Scribner's Sons, 1959), 351.

187 Spencer had a one-word term for the man: "fossilized": His comment came in an undated letter to Horace Cheney, MS79282.XXI, Cheney Family—Letters 1778–1933, Connecticut Historical Society, Hartford, CT.

188 "A great evil now specially prevalent": *The War of the Rebellion: A Compilation of the Official Records of the Union and Confederate Armies* (hereinafter OR), (Washington, DC: Government Printing Office, 1880–1901), Series III, Volume I, 264–65.

188 "As he has been for a long time making them": "A Revolving Patriot," was in the *New York Times*, April 26, 1861.

189 Chief of Ordnance Ripley liked enough to order: Ripley's order of Remington revolvers is quoted in Roy Marcot, *The History of Remington Firearms* (Guilford, CT: Lyons Press, 2005), 26.

189 "In manhood's strong and vigorous prime": Eliphalet Remington's deathbed poem and descriptions of his sons come in part from Hatch, *Remington Arms in American History,* 79–80.

190 "Go to church": Philo telling a workman with a weakness for alcohol to attend church is from Alden Hatch, *Remington Arms in American History* (USA: Remington Arms, 1856, rev. 1972), 80.

190 Samuel, the second Remington son: The Remington brothers' various ventures and attributes are discussed in Hatch, *Remington Arms in American History*, and in "Ilion and the Remingtons," an address given by Albert N. Russell before the Herkimer County Historical Society, September 14, 1897, *Papers Read Before the Herkimer County Historical Society During the Years 1896, 1897 and 1898*, vol. 1, pt. 1, compiled by Arthur T. Smith, Secretary of the Society (Herkimer, NY: Citizen, 1899), pp. 79–85 of the 1897 papers.

191 "the Government is already pledged on orders and contracts": Ripley's letter to Cameron on the Henry and Spencer rifles is in OR, Series III, Volume I, 733–734.

191 Charles Cheney had too much riding on the rifle: The negotiations that resulted in a contract for Spencer rifles are set forth in Marcot, *Spencer Repeating Firearms*, 29–35.

23: CATCHING UP TO WAR: 1862

193 "Spencer's famous breech loading rifle": "Improvement in Breech-Loading Rifles," *Scientific American* 6, no. 4 (January 25, 1862), 49.

193 The Chickering Piano-Forte Company: On pianos and the Chickering company, see Dale Tsang-Hall, "The Chickering Piano Company in the Nineteenth Century" (DMA thesis, Rice University, Houston, 2001).

193 "Piano-Fortes are becoming so fashionable": quoted in Gary J. Kornblith, "The Craftsman as Industrialist: Jonas Chickering and the Transformation of American Piano Making," *The Business History Review* 59, no. 3 (Autumn 1985): 353.

194 The eldest and senior partner: Col. Thomas E. Chickering advertised the bounty in the *Boston Evening Transcript*, September 17, 1862.

194 For Spencer that meant turning shafts: Accounts of Spencer's horseless carriage can be found in "The First Steam Car," *Hartford* (CT) *Daily Courant*, May 29, 1913; "Abraham Lincoln and the Repeating Rifle," *Scientific American* 125, no. 18 (December 1921): 102–3; and from a talk given by Spencer's daughter, Vesta Spencer Taylor, at a meeting of the Windsor Historical Society in September 1922, transcription at the Windsor Historical Society, Windsor, CT.

195 "Then he bade me a last farewell": Elizabeth Colt's account of her husband's last days is in Henry Barnard, *Armsmear: The Home, the Arm, and the Armory of Samuel Colt. A Memorial* (New York: Alvord, 1866), 323.

195 "an acute attack upon the brain: "Death of Col. Samuel Colt," *New York Times*, January 11, 1862.

195 Colt's funeral was as spectacular: A description of the funeral is in "The Funeral of Col. Sam Colt," *Hartford Daily* (CT) *Courant*, January 15, 1862.

196 Lincoln received many during the campaign and after: The vast array of presents given to Lincoln is discussed by Harold Holzer in "'Tokens of Respect' and 'Heartfelt Thanks': How Abraham Lincoln Coped with Presidential Gifts," *Illinois Historical Journal* 77, no. 3 (Autumn 1984): 177–92.

197 "I have heard of wars": quoted in Allen C. Guelzo, *Fateful Lightning: A New History of the Civil War & Reconstruction* (Oxford: Oxford University Press, 2012), 207.

197 Riding over the killing ground: Grant's quotes are from *Personal Memoirs of U. S. Grant,* vol. 1 (New York: Charles L. Webster, 1885), 356. His quote about realizing that only "complete conquest" would end the war is on p. 368.

198 "quite as cheap or cheaper than our own": Stuart C. Mowbray and Jennifer Heroux, eds., *Civil War Arms Makers and Their Contracts* (facsimile reprint of *The Report by the Commission on Ordnance and Ordnance Stores, 1862* (Lincoln, RI: Andrew Mowbray, 1998), 133.

198 Only about 125 guns: The production numbers of Henry rifles in 1862 are from Wiley Sword, *The Historic Henry Rifle: Oliver Winchester's Famous Civil War Repeater* (Lincoln, RI: Andrew Mowbray, 2002), Appendix A.

24: A BARON'S BID FOR DYNASTY

200 Winchester made sure to give his son: Information on William Wirt Winchester and Sarah Pardee comes in part from Mary Jo Ignoffo, *Captive of the Labyrinth: Sarah L. Winchester, Heiress to the Rifle Fortune* (Columbia: University

of Missouri Press, 2010); and Laura Trevelyan, *The Winchester: The Gun That Built an American Dynasty* (New Haven, CT: Yale University Press, 2016). William's description is in his application for a passport when he was twenty.

200 It was said he had "a retiring disposition": "William W. Winchester," *New Haven* (CT) *Register*, March 8, 1881.

201 Sarah's father, Leonard: Information about Leonard Pardee comes in part from Donald Lines Jacobus, ed., *The Pardee Genealogy* (New Haven, CT: New Haven Colony Historical Society, 1927), 244–45; and a Leonard Pardee & Co. advertisement in *The New England Business Directory* (Boston: Adams, Sampson, 1865), 187.

202 "Everything now is quiet": George Washington Whitman's letter to his mother is in the Trent Collection of Whitmaniana, Duke University Rare Book, Manuscript, and Special Collections Library.

202 Elsewhere in the crippled nation: For an account of the First Battle of Newtonia, see Larry Wood, *The Two Civil War Battles of Newtonia* (Charleston, SC: History Press, 2010).

25: THE HENRY OR THE SPENCER?

204 "We will not sell or furnish . . .": New Haven Arms Company Letters Book, p. 100, Winchester Repeating Arms Company Archives Collection, MS20.50.25, McCracken Research Library, Buffalo Bill Center of the West, Cody, WY.

204 "the most beautiful and efficient rifle we ever saw": "A Terrible Weapon," *Louisville* (KY) *Daily Journal*, June 26, 1862.

205 "the distinguished and patriotic journalist": excerpt from a letter Welles wrote in November 1861. Gideon Welles Correspondence, Vol. I, Box III. Folder 1, Connecticut Historical Society, Hartford, CT.

205 "Mothers named their children for him": Betty Carolyn Congleton, "George D. Prentice: Nineteenth Century Southern Editor," the *Register of the Kentucky Historical Society* 65, no. 2 (April 1967): 108.

205 "Please display & make the same attractive": excerpt from a letter on behalf of Oliver Winchester from J.H. Conklin to Prentice, March 3, 1863, New Haven Arms Company Letters Book, 188.

206 "This young man": Prentice's column in the *Journal* mourning his son's death was picked up by several other newspapers, including *The New York Times*, October 12, 1862.

206 "I am in bad spirits, my dear friend": John James Piatt, ed., *The Poems of George D. Prentice Edited with a Biographical Sketch* (Cincinnati, OH: Robert Clarke, 1876), 39.

206 "The women and children of this city": Nelson's evacuation report was printed

in the *Louisville Daily Journal*, September 23, 1862. His warning to stay at home or be shot comes from Bryan S. Bush, *Louisville and the Civil War: A History & Guide* (Charleston, SC: History Press, 2008), 56.

206 In 1857 he had shot it out: The Prentice duel is discussed in Berry Craig, "Old Time Kentucky: Louisville newspaper editors used pistols to take war of words to the next level," *Northern Kentucky Tribune,* March 5, 2016. www.nkytribune.com/2016/03/old-time-kentucky-louisville-newspaper-editors-used-pistols-to-take-war-of-words-to-the-next-level/.

207 "We have to beg that": New Haven Arms Company Letters Book, 132.

207 "We regret the mischief": New Haven Arms Company Letters Book, 126.

207 "Mr. Prentice, we have no doubt": New Haven Arms Company Letters Book, 110.

208 "The assets of the Company": New Haven Arms Company Letters Book, 118–20.

208 No Spencer repeating rifles: Rifle production details of the Spencer company can be found in Roy M. Marcot, *Spencer Repeating Firearms* (Rochester, NY: Rowe, 1983, 1990).

209 Spencer was a tightwad: Spencer's accounting of his expenses is cited in Marcot, *Spencer Repeating Firearms*, 51.

209 He already had one convert: Resource material on Wilder includes Stephen Cox, "Chattanooga Was His Town: The Life of General John T. Wilder," *Chattanooga Regional Historical Quarterly* 7, no. 1 (2004); Samuel C. Williams, "General John T. Wilder," *Indiana Magazine of History* 31, no. 3 (September 1935): 169–203.

209 "At what price will you furnish me": quoted in William B. Edwards, *Civil War Guns* (Harrisburg, PA: Stackpole, 1962), 161.

211 "Col. Wilder has just received": "Military Items," *Evansville (IN) Daily Journal*, May 23, 1863, quoting the *Indianapolis Journal*.

211 "We have drawn new guns: quoted in Dale Edward Linvill, ed., *Battles, Skirmishes, Events and Scenes: The Letters and Memorandum of Ambrose Remley* (Crawfordsville, IN: Montgomery County Historical Society, 1997), 58.

211 Single-shot muzzle-loaders: Muzzle-loaders being too slow for fast Americans is from B. F. McGee, *History of the 72d Indiana Volunteer Infantry of the Mounted Lightning Brigade* (Lafayette, IN: S. Vater, 1882), 120.

211 Wilder's few scouts refused to use Spencers: Andrew L. Bresnan, "Wilder's Brigade & Henry Rifles," *The Winchester Collector* (Summer 2014): 26–32.

211 "Such, however, is not the case": O.F.W., "Breech-loading versus Muzzle-loading Guns," *Scientific American* 8, no. 10 (March 7, 1863): 150–51.

212 "Will you obtain a Spencer rifle for us": New Haven Arms Company Letters Book, 211.

213 "Our Cartridge works blew up": New Haven Arms Company Letters Book, 303.

213 "[I]f these arms are used as efficiently": New Haven Arms Company Letters Book, 301.

214 As the Hatchet Brigade rode from camp": Accounts of the Battle of Hoover's Gap include Glenn W. Sunderland, *Lightning at Hoover's Gap: The Story of Wilder's Brigade* (New York: Thomas Yoseloff, 1969); Richard A. Baumgartner, *Blue Lightning: Wilder's Mounted Infantry Brigade in the Battle of Chickamauga* (Huntington, WV: Blue Acorn, 2007); and Robert S. Brandt, "Lightning and Rain in Middle Tennessee: The Campaign of June-July 1863," *Tennessee Historical Quarterly* 52, no. 3 (Fall 1993): 158–69. See also *The War of the Rebellion: A Compilation of the Official Records of the Union and Confederate Armies* (hereinafter OR), (Washington, DC: Government Printing Office, 1880–1901), Series I, Vol. XXIII, Part I, Chapter XXXV, 454–56; Confederate Brig. Gen. William B. Bate's report in OR, Series I, Volume XXIII, Chapter XXXV, Part 1, 611–14; William T. Alderson, "The Civil War Diary of Captain James Litton Cooper, September 30, 1861 to January, 1865," *Tennessee Historical Quarterly* 15, no. 2 (June 1956): 141, 155–56; and B. F. McGee, 128–32.

214 "[W]e knew full well, from the direction we were taking": *Three Years in the Army of the Cumberland: The Letters and Diary of Major James A. Connolly*, Paul M. Angle, ed. (Bloomington: Indiana University Press, 1959), 90.

215 "The boys said it was all 'Old Rosey'": excerpt from a letter written by a soldier in the First Wisconsin Infantry who called himself "Semi Occasional," published in the *Polk County* (WI) *Press*, August 22, 1863.

215 It was "no Presbyterian rain": quoted in David G. Moore, *William S. Rosecrans and the Union Victory: A Civil War Biography* (Jefferson, NC: McFarland, 2014), 86.

215 "was so much surprised by our sudden appearance": Connolly, *Three Years in the Army of the Cumberland*, 90.

216 "so close to us as to make it seem that the next would tear us to pieces": Connolly, 92.

217 "'Soldiers,' he proclaimed": quoted in W. J. McMurray, M.D., *History of the Twentieth Tennessee Regiment Volunteer Infantry, C.S.A.* (Nashville, TN: Publication Committee, 1904), 252.

217 "'Steady.' The word came from lip to lip along the line": Capt. John S. Wilson, quoted in Eric Nelsen Maurice, "'Send Forward Some Who Would Fight': How John T. Wilder and His 'Lightning Brigade' of Mounted Infantry Changed Warfare" (master's thesis, Department of History & Anthropology, Butler University, Indianapolis, 2016): 64.

218 "[B]ut few men": Connolly, 92.

218 "The effect of our terrible fire": *Sketches of War History, 1861–1865, Papers*

Prepared for the Commandery of the State of Ohio, Military Order of the Loyal Legion of the United States, 1903–1908, vol. 6 (Cincinnati, OH: Commandery of the State of Ohio, 1908), 173.

218 The Spencers' all-but-inexhaustible firepower: Repeating rifles' terrifying effect in combat may have been the result of their sound and fury more than their bullets. See Earl J. Hess, *The Rifle Musket in Civil War Combat: Reality and Myth* (Lawrence: University Press of Kansas, 2008), 52–59.

218 "as there was no danger of our being driven out of the position: *Sketches of War History,* 171.

218 "Thank God for your decision": *Sketches of War History,* 171–72.

218 "This engagement thoroughly tested the power of the Spencer rifles": excerpt from a letter Wilder wrote to Capt. George S. Wilson, who quoted it in "Wilder's Brigade: The Mounted Infantry in the Tullahoma-Chickamauga Campaigns," National (DC) Tribune, October 12, 1893.

26: GETTYSBURG

220 "To you I should have written long ago": Quotations from Noah Ferry's letter to his Aunt Mary come from Rev. David M. Cooper, *Obituary Discourse on Occasion of the Death of Noah Henry Ferry, Major of the Fifth Michigan Cavalry, Killed at Gettysburg, July 3, 1863* (New York: John F. Trow, 1863), 18.

220 They had been in Washington days earlier: The Michiganders' duties in Washington are mentioned in Jeffrey Alan Prins, "The Michigan Cavalry Brigade in the Civil War: A study of the tactical and strategic evolution of the Union cavalry" (master's thesis, Department of History, Central Michigan University, Mount Pleasant, 1994), 14.

221 A deep in his core abolitionist: Blair's activities on behalf of emancipation and enfranchisement are discussed throughout in Robert Charles Harris, "Austin Blair of Michigan: A Political Biography" (Ph.D. diss., Michigan State University, East Lansing, 1969). His 1855 legislation is mentioned on p. 68.

221 "slavery should be swept from the land": Joint Resolution No. 12, *Acts of the Legislature of the State of Michigan, Passed at the Extra Session of 1862,* 67–68.

221 Impatient for the Lincoln administration: Blair's impatience is mentioned in Jean Joy L. Fennimore, "Austin Blair: Civil War Governor, 1861–1862," *Michigan History* 49 (September 1965): 205.

221 When the Spencers finally did show up: The saga of getting Spencer rifles to the Michigan regiments is detailed in Wiley Sword, "'Those Damned Michigan Spencers,'" *Man at Arms* 19, no. 5 (October 1997): 23–37.

221 "At last we are soldiers": excerpt from a letter written by Victor Comte on January 6, 1863. Victor E. Comte papers: 1853–1878, Bentley Historical Library, University of Michigan, Ann Arbor.

222 Noah Ferry fit that profile: Discussion of the life Ferry left to join the Army is from J. H. Kidd, *Personal Recollections of a Cavalryman with Custer's Michigan Cavalry Brigade in the Civil War* (Ionia, MI: Sentinel, 1908), 108–9. Other information about Ferry comes from Cooper, *Obituary Discourse*, and Thomas Holbrook, "Men of Action: The Unsung Heroes of East Cavalry Field" in "Unsung Heroes of Gettysburg," *Gettysburg Seminar Papers*, published by National Park Service, 1996. http://npshistory.com/series/symposia/gettysburg_seminars/5/essay5.htm.

222 He wore a brocaded black velvet jacket: Description of Custer's uniform and "turned down on one side" is from Kidd, *Personal Recollections of a Cavalryman*, 129–30; and T. J. Stiles, *Custer's Trials: A Life on the Frontier of a New America* (New York: Knopf, 2015), 94.

222 "one of the funniest-looking beings you ever saw": *Meade's Headquarters: Letters of Colonel Theodore Lyman from the Wilderness to Appomattox*, George R. Agassiz, ed. (Boston: Atlantic Monthly Press, 1922), 17.

223 "Come let us lie down and get a little sleep": quoted in Eric J. Wittenberg, *Protecting the Flank at Gettysburg: The Battles for Brinkerhoff's Ridge and East Cavalry Field, July 2–3, 1863* (El Dorado Hills, CA: Savas Beatie, 2013), 41, quoting *The Bachelder Papers: Gettysburg in Their Own Words*, 3 vols. (Dayton, OH: Morningside Bookshop, 1994–95).

224 "Now, boys": Quotes from the battle are from Cooper, *Obituary Discourse*, 20, 22.

225 "with his yellow locks flying": Colonel Hampton S. Thomas, *Some Personal Reminiscences of Service in the Cavalry of the Army of the Potomac* (Philadelphia: L. R. Hamersly, 1889), 13.

225 In his report eight weeks after the battle: Custer's praise of the Spencer rifle comes from his August 22, 1863, report of the Battle on the East Cavalry Field, reproduced in Jno. Robertson, *Michigan in the War*, rev. ed. (Lansing, MI: W. S. George, 1882), 583.

225 Technically the fight there ended in a draw: Whether the battle at East Cavalry Field was of great consequence is debatable. Some say that the Union forces there, including Custer's troopers, thwarted Stuart's plan to get behind or beside the main Union Army as the famous Pickett's Charge slammed into the front. Historian Allen C. Guelzo says that "both sides drew off [after the battle] with little to show for it except for some minor casualties . . . except for Custer's brigade (where 32 men were killed and 147 wounded, a pattern of heedless bloodletting which Custer would carry to a more famous spot on a dusty hillside in Montana thirteen years later)." *Gettysburg: The Last Invasion* (New York: Knopf, 2013), 430.

27: A LIVING SHEET OF FLAME

226 "The result is that I have tried": Lincoln's complaints about the Spencer rifle come from an August 4, 1863, letter he wrote to General Hurlbut. The letter is in the collection of the Gilder Lehrman Institute of American History, New York, NY.

227–228 "of some originality of character" and "Lincoln had a quick comprehension of mechanical principles": *Addresses of John Hay* (New York, NY: Century, 1906), 327.

227 Lincoln also liked new things: Jason Emerson discusses Lincoln's fascination with mechanics in *Lincoln the Inventor* (Carbondale: Southern Illinois University Press, 2009).

227 As a lawyer, he took on a number of patent cases: On Lincoln's patent llitigation, see Jeffrey M. Samuels and Linda B. Samuels, "Lincoln and the Patent System: Inventor, Lawyer, Orator, President," *Albany Government Law Review* 3 (2010): 645–674; and Harry Goldsmith, "Abraham Lincoln, Invention and Patents," *Journal of the Patent Office Society* 20, no. 1 (January 1938): 5–33.

227 "It's like some of the glib talkers you and I know": Don Davenport, *In Lincoln's Footsteps: A Historical Guide to the Lincoln Sites in Illinois, Indiana, and Kentucky* (Madison, WI: Prairie Oak Press, 1991), 136.

227 in 1849 Lincoln patented a system of inflatable bellows: Some have criticized Lincoln's patent as unworkable. An engineer who studied it recently concluded that, while it had serious flaws, "it was a prescient concept and one that was scientifically tenable." Ian de Silva, "Evaluating Lincoln's Patented Invention," *Journal of the Abraham Lincoln Association* 39, no. 2 (Summer 2018): 1–28.

227 "in anciently inhabited countries": For Lincoln's Second Lecture on Discoveries and Inventions, see *Collected Works of Abraham Lincoln*, vol. 3 (Ann Arbor: University of Michigan Digital Library Production Services, 2001): 356–63. https://quod.lib.umich.edu/cgi/t/text/text-idx?c=lincoln;idno=lincoln3.

228 Chief of Ordnance General James Wolfe Ripley: Ripley's unwavering opposition to innovative small arms finally did him in. On September 15, 1863, he was removed as chief of ordnance and assigned to inspect fortifications along the New England coast, which he did for the rest of the war and beyond.

228 "and had ideas of his own": William O. Stoddard, ed. by William O. Stoddard, Jr., "Face to Face with Lincoln," *Atlantic Monthly* (March 1925): 335.

228 "looked like a gunshop": Ibid.

228 Lincoln would occasionally test guns: *Addresses of John Hay*, 327–28.

229 Perhaps Lincoln could be convinced: Warren Fisher's letter to Lincoln is in the Abraham Lincoln Papers, Series 1, General Correspondence, 1833–1916, Library of Congress, Washington, DC.

229 Christopher Spencer showed up at the White House: Spencer's account of his shooting with Lincoln comes from the Christopher Miner Spencer Collection, Series I.A, Correspondence, Box 1.2, Windsor Historical Society, Windsor, CT.

230 John Hay called Spencer's repeater a "wonderful gun": Hay's comments on Spencer and his rifle are in his diary entry of August 19, 1863. *Letters of John Hay and Extracts From Diary*, vol. 1 (Washington, DC, 1908): 93.

231 with Capt. Eli Lilly also shooting at them: A decade after the war Lilly started a pharmaceutical company bearing his name, which is still in existence today.

231 "Our entire line from right to left": the journal of Cpl. William H. Records of the Seventeenth Indiana, Manuscript Section, Indiana Division, Indiana State Library, Indianapolis, IN, quoted in Richard A. Baumgartner, *Blue Lightning: Wilder's Mounted Infantry Brigade in the Battle of Chickamauga* (Huntington, WV: Blue Acorn, 2007), 227.

232 "Again and again": Theodore Petzoldt, Seventeenth Indiana Mounted Infantry, *My War Story* (Portland, OR, 1917), 103–5, quoted in William Glenn Robertson, Edward P. Shanahan, John I. Boxberger, and George E. Knapp, *Staff Ride Handbook For The Battle Of Chickamauga, 18–20 September 1863* (Fort Leavenworth, KS: Combat Studies Institute, US Army Command and General Staff College, 1992), 89.

232 "The effect was awful": "Something More about the Battle of Chickamauga," *Indianapolis Daily Journal*, September 28, 1863.

232 "At this point": "The Battle of Chickamauga; Important and Interesting Statement By Col. Wilder," *New York Times*, October 4, 1863

232 "At a distance of less than fifty yards": James Burns, Thirty-Ninth Indiana Mounted Infantry, quoted in www.civilwarhome.com/hillatchickamauga.htm.

233 "such slaughter and carnage": B. F. McGee, *History of the 72d Indiana Volunteer Infantry of the Mounted Lightning Brigade* (Lafayette, IN: S. Vater, 1882), 176.

233 "How I wish our infantry": Private Alva Griest, quoted in Baumgartner, *Blue Lightning*, 304.

233 "We got through the three days of fighting": Major Connolly wrote the letter to his wife on October 21, 1863, from Chattanooga. It is printed in *Three Years in the Army of the Cumberland: The Letters and Diary of Major James A. Connolly*, Paul M. Angle, ed. (Bloomington: Indiana University Press, 1959, renewed 1987), 126–30.

28: NO BETTER FRIEND, NO WORSE ENEMY

234 "For God's sake": Wilson's account of his battle with guerrillas and his response to Winchester's letter are in H. W. S. Cleveland, *Hints to Riflemen* (New York: D. Appleton, 1864), 180–82.

235 "A few days since I heard": Winchester's letter to Wilson is in the New Haven Arms Company Letters Book, p. 175, Winchester Repeating Arms Company Archives Collection, MS20.50.25, McCracken Research Library, Buffalo Bill Center of the West, Cody, WY.

235 "Give us anything": excerpt from a letter Cloudman wrote to Oliver Winchester on March 15, 1865. Winchester printed the letter in the New Haven Arms Company's 1865 catalogue.

235 president of the strategically located: The Louisville and Nashville Railroad plea for repeating rifles is in a July 22, 1864, letter from J. Holt to Stanton, *The War of the Rebellion: A Compilation of the Official Records of the Union and Confederate Armies* (hereinafter OR) (Washington, DC: Government Printing Office, 1880–1901) Series I, Vol. XXXIX, Part II, 198.

236 "could cause terrific losses": The L.W.W. letter is quoted and discussed in George Madis, *The Winchester Era* (Brownsboro, TX: Art and Reference House, 1984), 43.

236 "An Easy Way to Avoid the Conscription": the *Gallipolis* (OH) *Journal*, March 26, 1863.

237 "From where I fell wounded": Mack Leaming's account of the Fort Pillow massacre is at the Gilder Lehrman Institute of American History, New York, NY.

237 "with their eyes punched out": The mate who saw the bodies of soldiers with their eyes bayonetted was Robert S. Critchell, Acting-Master's Mate, U.S.N. His report was in "The Fort Pillow Massacre: A Letter from a Naval Officer," *New York Times*, May 3, 1864.

237 Unreliable old guns might be fine: Hinks's letter to Butler is in OR, Series I, Vol. 33, 1020–21. For Confederate treatment of African American soldiers who fought for the Union, see Drew Gilpin Faust, *This Republic of Suffering: Death and the American Civil War* (New York: Knopf, 2008), 44–47.

238 Winchester demanded that damages be paid in the form of machinery: Winchester's Machiavellian move to force the Burnside company to stop making guns is discussed in Herbert G. Houze, *Winchester Repeating Arms Company: Its History & Development from 1865 to 1981* (Iola, WI: Krause, 1994, 2004), 63.

238 Although Henry kept things going: Information about Henry not increasing production and Winchester leasing a Bridgeport factory is from Houze, *Winchester Repeating Arms Company,* 18–19.

239 In 1859 White, along with Smith & Wesson: The suit Rollin White and Smith & Wesson brought against Ethan Allen is *Rollin White et al v. Ethan Allen et al,* Circuit Court of the United States, Massachusetts District (Boston: Alfred Mudge & Son, 1863).

240 "stop all purchases of arms": The order for the government to stop buying guns

is General Orders, No. 77, April 28, 1865, OR, Series III, Volume IV, Section 2, 1280–81.

241 In 1865 alone the firm sold 55,543 of them: The number of revolvers Smith & Wesson made in 1865 comes from Roy Jinks's files, as does the October 16, 1865, order book in which Daniel Baird Wesson wrote, "Very great are the changes. . . ."

29: THE WAGES OF PEACE

242 "the most fantastic": William B. Edwards, *Civil War Guns* (Harrisburg, PA: Stackpole, 1962), 400.

242 "[T]he armories are disbanding": "The War Ended," *Scientific American* 12, no. 26 (June 24, 1865): 407.

242 Without a network of private companies: The need for a network of private firms to meet war needs is discussed in Ross Thomson, "The Continuity of Innovation: The Civil War Experience," *Enterprise & Society* 11, no. 1 (March 2010): 135. "In total, private domestic firms doubled the Armory's output over the course of the war and introduced new types of military firearms, clearly a remarkable accomplishment." Thomson, "The Continuity of Innovation . . . ," 134.

243 In the five years following the Civil War: On the gun companies postwar, see Felicia Johnson Deyrup, *Arms Makers of the Connecticut Valley: A Regional Study of the Economic Development of the Small Arms Industry, 1798–1870* (Menasha, WI: George Banta, 1948), 202–16.

243 Before the Civil War's end: Information on Oliver Winchester after the war and the souring of relations between him and B. Tyler Henry, including the quote from Winchester's cable to Davies, comes from Herbert G. Houze, *Winchester Repeating Arms Company: Its History & Development from 1865 to 1981* (Iola, WI: Krause, 1994 and 2004), 18–25; and Dean K., Boorman, *The History of Winchester Firearms* (Guilford, CT: Lyons Press, 2001), 20–21, 24.

246 Winchester's wealth and his readiness "to respond liberally to calls made for party purposes": "The Connecticut Republican Convention," Washington *Evening Union*, February 16, 1866.

246 "The manner in which Mr. Winchester received the nomination": "Connecticut—How He Got the Nomination," *Boston Post*, March 29, 1866.

247 conservatives whose Southern "sympathies and affinities": "A Disgraceful Affair," *Hartford* (CT) *Daily Post*, April 19, 1866.

248 "my business relations": Winchester's letter declining for business reasons to run again and attacking the Radicals appeared in several newspapers in the country, as well as those in Connecticut. In New Haven it was printed in the *Columbian Register*, February 16, 1867.

249 "In *range* and *force*" the Spencer Rifle is second to no other arm": the Spencer Repeating Rifle Company's 1866 catalogue, 2–3.

249 The war has nearly settled the question": Grant's encomium for the Spencer rifle is from a letter he wrote to Warren Fisher, Jr., on January 12, 1866, printed on page 27 of the Spencer Repeating Rifle Company's 1866 catalogue.

250 "More rapid firing than this": the 1866 Spencer catalogue, 1.

251 he auctioned off all the equipment: A report detailing the Spencer company auction was "Large Auction Sale: The Spencer Repeating Rifle Works Disposed of—Total Proceeds $138,000," *Boston Post*, September 29, 1869.

251 After Spencer returned from Amherst: The 1870 census shows Dora and her sister living with their parents. The 1880 census gives information on Sardis Peck's mental state.

251 Affable salesman and brother: Samuel Remington and his exploits in Europe and Egypt are discussed in Alden Hatch, *Remington Arms in American History*, rev. ed. (USA: Remington Arms, 1972), 142–45; K.D. Kirkland, *America's Premier Gunmakers: Remington* (New York: Exeter Books, 1988), 36–38; John P. Dunn, *Khedive Ismail's Army* (London: Routledge, 2005), 65–70; and John Dunn, "Remington 'Rolling Blocks' in the Horn of Africa," *American Society of Arms Collectors Bulletin*, no. 71 (Fall 1994): 20–21.

252 Later dubbed the "rolling block": For information on the rolling block, see George J. Layman, *The Military Remington Rolling Block—50 Years of Faithful Service* (Prescott, AZ: Wolfe, 1992), and *The All New Collector's Guide to Remington Rolling Block Military Rifles of the World* (Woonsocket, RI: Andrew Mowbray, 2010); Marcot, *Remington: "America's Oldest Gunmaker"* (Peoria, IL: Primedia, 1998), Chapter 4; Dunn, "Remington 'Rolling Blocks' in the Horn of Africa," 19–31; and Dunn, *Khedive Ismail's Army.*

253 "there is no North or South here": excerpt from a letter written by Col. Charles I. Graves to his wife, quoted in Dunn, "Remington 'Rolling Blocks' in the Horn of Africa," 21.

254 Between September 21, 1870, and May 6, 1871: Statistics on the number of guns shipped from the Remington factory and the quote "not surprised that" come from the *Army and Navy Journal* 8, no. 38 (May 6, 1871): 604.

255 "I should rather be the man to be shot at": excerpt from the deposition of Luke Wheelock in *Rollin White et al v. Ethan Allen et al*, Circuit Court of the United States, Massachusetts District (Boston: Alfred Mudge & Son, 1863), 99.

256 Despite White's poormouthing: Dyer's letter and Grant's veto are in "Message of the President of the United States Returning Bill S. No. 273, entitled 'An act for the relief of Rollin White,' with his objections," January 11, 1870, 41st Congress, 2nd Session, Ex. Doc. No. 23.

256 With stunning vitriol: Quotes from the Remonstrants' brief are from Dodge

& Son, *Reasons Why the Bill for the Relief of Rollin White Ought Not To Pass* (Washington, DC: S. & R. O. Polkinhorn, 1874), 11, 26, 34.

258 George had ambition as well as a gift for mechanics: The letter from Schofield to Wesson is in Roy G. Jinks, *History of Smith & Wesson: No Thing of Importance Will Come Without Effort,* 14th printing (North Hollywood, CA: Beinfeld, 2004), 86.

260 Nevertheless, Schofield eventually became: The reports of Schofield's career and suicide come from Constance Wynn Altshuler, *Cavalry Yellow & Infantry Blue* (Tucson: Arizona Historical Society, 1991), 293–94; "By His Own Hand," *Streator* (IL) *Free Press,* December 23, 1882; "Recent Deaths—Brevet Colonel G. W. Schofield, U.S.A.," *Army and Navy Journal* 20, no. 21 (December 23, 1882): 461; and Donald B. Connelly, *John M. Schofield and the Politics of Generalship* (Chapel Hill: University of North Carolina Press, 2006), 258–59.

30: CENTENNIAL

261 The monster machine: Authors Nathan Rosenberg and Manuel Traitenberg credit the Corliss system with playing a "key role in the fierce contest between waterpower and steam power, particularly in the Northeast. In so doing it helped propel the steam engine to a dominant position in the intertwined processes of industrialization and urbanization that characterized the growth of the U.S. economy in the second half of the nineteenth century." "A General-Purpose Technology at Work: The Corliss Steam Engine in the Late-Nineteenth-Century United States," *Journal of Economic History* 64, no. 1 (March 2004): 62.

262 "There it stands": Phillip T. Sandhurst and Others, *The Great Centennial Exhibition Critically Described and Illustrated* (Philadelphia: P. W. Ziegler, 1876), 361.

263 "a most interesting exhibit of asbestos": James D. McCabe, *The Illustrated History of the Centennial Exhibition* (Philadelphia: Jones Brothers, 1876), 449.

263 "the pop-corn man had a tasteful stand": Information about popcorn and the fees for selling it comes from McCabe, *The Illustrated History,* 437, 640. The "popcorn capitalist" quote is from "The Great Exhibition: One of Its Peculiar Features Explained," *New York Daily Herald,* February 23, 1876.

263 "The negroes, as they work": McCabe, 452.

263 The colossal forearm and torch-bearing hand: The quote about Liberty's commodious hand is from "The French Statue," *New York Times,* September 29, 1876.

263 "Almost every country showed arms": Nancy Kizer, "The Story of the Philadelphia Centennial of 1876," *Mississippi Quarterly* 7, no. 3 (April 1954): 19.

264 "all sorts of Guns": William Randel, "John Lewis Reports the Centennial," *Pennsylvania Magazine of History and Biography,* July 1955, 370.

264 "A study of the savage weapons at the Centennial exhibition, Philadelphia, 1876": Edward H. Knight is in the *General Appendix to the Smithsonian Report for 1879.*

264 In the government building the Ordnance Department: The cartridge-making display is discussed in Berkeley R. Lewis, *Small Arms Ammunition at the International Exposition Philadelphia, 1876* (Washington, DC: Smithsonian Institution Press, 1972), 9, 49.

264 There "the skilful men operatives": McCabe, *The Illustrated History,* 582.

264 "There were pistols and revolvers enough to arm the Russian soldiery": J. S. Ingram, *The Centennial Exposition, Described and Illustrated* (Philadelphia: Hubbard Bros., 1876), 124.

264 More than three hundred revolvers: The Colt display is described in Carol Wilkerson, "The 1876 Centennial Exhibition in Philadelphia, Pennsylvania," *The Rampant Colt,* Spring 2014, 46; and George D. Curtis, *Souvenir of the Centennial Exhibition: Or, Connecticut's Representation at Philadelphia, 1876* (Hartford, CT: George D. Curtis, 1877), 83.

265 For its part the Winchester company: The Winchester display is described in Curtis, *Souvenir of the Centennial Exhibition,* 88.

265 "as handsome specimens of work and finish": Comments on the Remington exhibit come from *Frank Leslie's Illustrated News,* July 15, 1876.

266 exhibit to "illustrate the past and present condition": Baird's motives and quotes are in Judy Braun Zegas, "North American Indian Exhibit at the Centennial Exposition," *Curator: The Museum Journal,* 19, no. 2 (June 1976): 163.

266 "The presence of a band of uncivilized aborigines": several papers published this, including the *National Republican* (Washington, DC), August 7, 1875.

266 "ready to pounce": quoted in Robert A. Trennert, "Popular Imagery and the American Indian: A Centennial View," *New Mexico Historical Review* 51, no. 3 (July 1976): 223–24.

267 "It is certainly curious": "Centennial Notes," *Philadelphia Inquirer,* July 26, 1876.

31: SACRED IRON

268 A Sioux term for a gun was "sacred iron," because it had been thought to possess supernatural power. See introduction by John C. Ewers in Frank Raymond Secoy, *Changing Military Patterns of the Great Plains Indians* (Lincoln: University of Nebraska Press, 1992), xii. (Originally published by the University of Washington Press, 1953, in the series, Monographs of the American Ethnological Society).

268 Lt. Col. George Armstrong Custer: After the Civil War ended, Custer's rank of brevet major general was reduced to lieutenant colonel, as was customary

for officers staying in service when the need for a large military force was over. People still addressed him as "general."

269 "You speak of another country": The Spotted Tail quote comes from Pekka Hämäläinen, *Lakota America: A New History of Indigenous Power* (New Haven, CT: Yale University Press, 2019), 349.

269 Plains Indians had long been gathering weapons: The role of guns as symbols of Indian manhood comes from David J. Silverman, *Thundersticks: Firearms and the Violent Transformation of Native America* (Cambridge, MA: Belknap Press of Harvard University Press, 2016), 9.

269 A horse or a mule could pay for a repeating rifle: Trading animals and buffalo hides for guns and ammunition comes from James Donovan, *A Terrible Glory: Custer and the Little Bighorn—the Last Great Battle of the American West* (New York: Little, Brown, 2008), 188.

269 "The Sioux must have good white men friends on the Platte and Missouri": *Fifth Annual Report of the Board of Indian Commissioners to the President of the United States. 1873* (Washington, DC: Government Printing Office, 1874), 105. Some referred to the Springfield as a "needle gun" because of its long firing pin between the hammer and the cartridge.

269 after a skirmish with the Crows' longstanding Indian enemy: "The Crow, Arikara, and other tribes had been fighting the Sioux for generations. . . . They still suffered from Sioux aggression during the 1860s and 1870s." John C. Ewers, "Intertribal Warfare as the Precursor of Indian-White Warfare on the Northern Great Plains," *Western Historical Quarterly* 6, no. 4 (October 1975): 409–10.

269 "The arms with which they fought us": Custer's 1873 report on well-armed Indians is reproduced in Elizabeth B. Custer, *Boots and Saddles: or Life in Dakota with General Custer* (New York: Harper & Brothers, 1885), 268–79.

270 Now, with Winchesters and Spencers in their arsenal: The role of the Winchester '66 in helping to revolutionize plains warfare is mentioned in Hämäläinen, *Lakota America*, 299. Intertribal arms races are mentioned in Silverman, *Thundersticks*, 8.

270 that transformed "the Plains Indian from an insignificant, scarcely dangerous": Richard Irving Dodge, *Our Wild Indians: Thirty-Three Years' Personal Experience Among the Red Men of the Great West* (Hartford, CT: A. D. Worthington, 1883), 450–51.

271 Custer was a good shot: Custer's buffalo hunt and the inadvertent shooting of his horse are from Gen. G. A. Custer, *My Life on the Plains. Or, Personal Experiences with Indians* (New York: Sheldon, 1874), 37–38.

271 "but a portion of the game killed by me": Custer's October 5, 1873, letter to the Remington company is printed in John S. du Mont, *Custer Battle Guns* (Fort Collins, CO: Old Army Press, 1974), 68.

271 "Sitting Bull, thy doom is fast approaching": a letter to the editor by Long Horse, "The Indian Campaign," *Helena* (MT) *Weekly Herald*, June 29, 1876, before word of Custer's defeat had spread.

272 Oglala warrior Eagle Elk: Description of Eagle Elk attacking Reno's troopers is from David Humphreys Miller, *Custer's Fall: The Native American Side of the Story* (New York: Meridian, 1992), 109.

273 "the snapping of the threads in the tearing of a blanket": Crow scout Curley, quoted in Nathaniel Philbrick, *The Last Stand: Custer, Sitting Bull, and the Battle of the Little Bighorn* (New York: Viking, 2010), 267.

273 "We advise the Winchester Arms Company": the *Army and Navy Journal* 13, no. 50 (July 22, 1876): 805.

274 It was numbers, tactics, and fearless commitment to a way of life: On superior Indian tactics and strategy, see Bradley C. Vickers, "More Than Numbers: Native American Actions at the Battle of the Little Bighorn" (master's thesis, United States Marine Corps, Command and Staff College, Marine Corps University, Quantico, VA, 2002).

EPILOGUE

277 "Mr. White claimed the invention": The obituary for Rollin White was in the *Rutland* (VT) *Daily Herald*, March 28, 1892.

277 "He was, just before his last illness": B. Tyler Henry's obituary was in the *Daily Morning Journal and Courier* (New Haven, CT), June 8, 1898.

280 "I'd always hoped to leave you money": "Character Sketch of Christopher Miner Spencer, Written by his daughter, Vesta Spencer Taylor," Spencer Collection, Series II, Box 1.11, Windsor Historical Society, Windsor, CT.

INDEX